Lecture Notes in Physics

Volume 974

The series Lecture Notes in Physics (LNP), founded in 1969, reports new developments in physics research and teaching - quickly and informally, but with a high quality and the explicit aim to summarize and communicate current knowledge in an accessible way. Books published in this series are conceived as bridging material between advanced graduate textbooks and the forefront of research and to serve three purposes:

- to be a compact and modern up-to-date source of reference on a well-defined topic.
- to serve as an accessible introduction to the field to postgraduate students and nonspecialist researchers from related areas.
- to be a source of advanced teaching material for specialized seminars, courses and schools.

Both monographs and multi-author volumes will be considered for publication. Edited volumes should however consist of a very limited number of contributions only. Proceedings will not be considered for LNP.

Volumes published in LNP are disseminated both in print and in electronic formats, the electronic archive being available at springerlink.com. The series content is indexed, abstracted and referenced by many abstracting and information services, bibliographic networks, subscription agencies, library networks, and consortia.

Proposals should be sent to a member of the Editorial Board, or directly to the managing editor at Springer:

Dr Lisa Scalone
Springer Nature
Physics Editorial Department
Tiergartenstrasse 17
69121 Heidelberg, Germany
lisa.scalone@springernature.com

More information about this series at http://www.springer.com/series/5304

Osvaldo Civitarese • Manuel Gadella

Methods in Statistical Mechanics

A Modern View

 Springer

Osvaldo Civitarese
Department of Physics
University of La Plata
La Plata, Buenos Aires, Argentina

Manuel Gadella (iD)
Department of Theoretical Physics
Atomic Physics and Optics
University of Valladolid
Valladolid, Spain

ISSN 0075-8450   ISSN 1616-6361   (electronic)
Lecture Notes in Physics
ISBN 978-3-030-53657-2   ISBN 978-3-030-53658-9   (eBook)
https://doi.org/10.1007/978-3-030-53658-9

This Springer imprint is published by the registered company Springer Nature Switzerland AG
The registered company address is: Gewerbestrasse 11, 6330 Cham, Switzerland

Preface

Among the various branches of theoretical physics, the statistical mechanics is present in the analysis of practically all physical systems, independently of the scale of them. Examples on the applications of statistical concepts are, for instance, the theory of massive astrophysical objects, the models of elementary particle systems at finite densities and temperatures, diverse types of classical and quantum fluids, the thermodynamics of real and ideal gases, etc. To these examples, we may add the extremely rich field of phase transitions, both quantum and classical.

The most widely adopted models of hadrons and their interactions, of nucleon and nuclear structure, atoms and molecules, and ultimately of extended bodies (say, e.g., solids, liquids, gases) often rely on concepts like densities, temperature and various forms of statistical equilibrium.

The conventional presentations of statistical mechanics usually resort to ideal systems without interactions, or to approximations like mean-field scenarios. To concepts like the statistical equilibrium and the correspondence between calculated and observable quantities, it would be desirable to add ingredients like interactions, finite size effects, dimensional dependence, symmetry breaking and symmetry restoring effects. Real systems are far to belong to the class of ideal ones, and their number of components are not always infinity, meaning that the vast domain of few particle systems constitutes a terra ignota from the point of view of statistical mechanics.

In this book, we are presenting different techniques meant to tackle some of the features exceeding the conventional approach. Consequently, we shall introduce methods based on the use of path integration, thermal Green functions, time-temperature propagators, Liouville operators, second quantization, and field correlators at finite density and temperature. We shall also address the question of the statistical mechanics of unstable quantum systems.

In writing this book, we have benefited from the existing literature. In selecting the examples about the applications of the various techniques, we have revisited some of the most influential books and papers in the field, as indicated in the text.

The following is a list of the contents of the book: Chap. 1 contains a revision of the classical and quantum statistical mechanics formulated for discrete and continuous spectra, based on the notion of probabilities and thermal equilibrium. Chapter 2 is devoted to the role of dynamics, with an emphasis on the connections between Liouville dynamics and statistical mechanics. Chapter 3 contains the notions of operators and their role in statistical mechanics, particularly in the combined time-temperature representations. Chapter 4 reviews the Feynman path-integral formulation. Chapter 5 explores the principles of statistical mechanics in terms of geometry. Chapter 6 is a formal continuation of the previous chapter, which specializes in the connections with the statistical ensembles. Chapter 7 contains the basic notions which support the use of statistical mechanics for unstable systems.

This book is meant to be used for a semester course, following graduate lectures in quantum mechanics, thermodynamics, electromagnetism and mathematical methods in physics. The material of the book is self-contained from the mathematical point of view, and the subjects are arranged sequentially.

The authors acknowledge the support of the National Research Council of Argentina (grant PIP-616), and of the Ministry of Economy and Productivity of Spain (grant MTM 2014-57129-C2-1-P) and of the Junta de Castilla y León (grant BU229P18). The hospitality of the Department of Physics (National University of La Plata, Argentina), the Institute of Physics of La Plata (National Research Council of Argentina), and of the Department of Atomic and Theoretical Physics and Optics of the University of Valladolid, Spain, where parts of this book have been written, is gratefully acknowledged by the authors.

La Plata, Argentina   Osvaldo Civitarese
June 2020   Manuel Gadella

Contents

About the Authors

 **Dr. Osvaldo Civitarese** is an Emeritus Professor of Theoretical Physics at the University of La Plata, Argentina. He is an expert in nuclear structure and neutrino physics. During his tenure, he gave lectures in statistical mechanics, electromagnetism, quantum mechanics and nuclear structure. He has published more than 300 research papers, 50 conference articles and 1 book on modern physics for undergraduate students. He got national and international awards like the J. S. Guggenheim Foundation prize. He is a member of the Mexican Academy of Science and a former Alexander von Humboldt fellow.

 **Dr. Manuel Gadella** is a Professor of Physics at the University of Valladolid, Spain. He is an expert in mathematical physics. He is author of about 150 research papers and 2 books for undergraduate students. In collaboration with Prof. Arno Bohm (University of Texas at Austin), he wrote the book entitled "Dirac kets, Gamow vectors and Gelfand triplets" published in the Springer Lecture Notes in Physics.

An Introduction to Statistical Mechanics

In this chapter, we are going to present the essentials of the probabilistic formulation of the statistical mechanics. We shall follow the Gibbs approach [1–3] based on the notion of equilibrium. This formulation applies to both classical and quantum systems. To illustrate the concept of probability, we shall start with a system with discrete energy levels. Then, by applying a limiting process to this discrete spectrum, we arrive at a classical statistical description. For historical reasons, we shall first introduce the concept of partition function, and then proceed to the definition of statistical averages and discuss the connections with thermodynamical potentials and their associated observables. Next, we shall work with quantum representations and compare the classical distribution with the high temperature limit of their quantum counterparts. The material included in the present chapter is mathematically self-contained.

1.1 Partition Functions

The conventional presentation of statistical mechanics relies upon the notion of probabilities, which are defined in terms of the constraints imposed on the system, like fixed energy, number of particles, etc. As explained in practically all books dealing with statistical mechanics, with few exceptions, one should speak in terms of statistical ensembles. In each of them (Micro-canonical, Canonical, Grand Canonical), the starting point is just the definition of the energy states.which are accessible to the components of the system. After it, one should also rely upon the notion of statistical equilibrium depending upon the values of extensive and intensive variables. To this one should add fluctuations and eventually associate them with the rates of heat or

O. Civitarese and M. Gadella, *Methods in Statistical Mechanics*, Lecture Notes in Physics 974, https://doi.org/10.1007/978-3-030-53658-9_1

energy transfer from the system to the environment. In this chapter, we shall revisit these notions starting from the distinction between discrete and continuum spectra, either classical or quantum.

1.1.1 Systems with Discrete Spectrum

Let us consider a system for which the dynamics is governed by a Hamiltonian H, with a purely non-degenerate discrete spectrum. This will be the case of N particles confined in a box of fixed volume. The set of real values, $\{E_1, E_2, \ldots, E_r, \ldots\}$, is the spectrum of H. The probability, p_r that the system be in the state E_r, in thermal equilibrium with the bath at temperature T, is $p_r = Z^{-1} e^{-\beta E_r}$, where Z is a normalization constant, $\beta = (k_B T)^{-1}$ and k_B is the Boltzmann constant ($k_B = 1.3806488 \times 10^{-23}\, J K^{-1}$). The hypothesis of equilibrium relies upon the a priori separation between the system under study and the environment (i.e., everything save for the system). This separation is based on the notion that changes in the system, induced by the interaction with the environment, are much larger than the changes experimented by the environment. In general, the number of states per unit energy, in most known systems, increases exponentially with energy. Then, the level density may be represented by the form

$$\rho(E) \approx \rho_0\, e^{\beta E} . \tag{1.1}$$

At first order the variation $\delta\rho$ is written as

$$\beta = \frac{1}{\rho}\frac{\delta\rho}{\delta E} , \tag{1.2}$$

which constitutes a definition of the temperature.

The sum of these probabilities must be equal to one, $\sum_r p_r = 1$; then,

$$p_r = \frac{e^{-\beta E_r}}{\sum_r e^{-\beta E_r}} . \tag{1.3}$$

The *partition function* Z is defined as the sum in the denominator in (1.3):

$$Z = \sum_r e^{-\beta E_r} . \tag{1.4}$$

If some of the energy levels E_r is degenerate, we shall take into account this degeneracy in order to construct the partition function, so that in general, we have

$$Z = \sum_r g(E_r)\, e^{-\beta E_r} , \tag{1.5}$$

where $g(E_r)$ is the degeneracy corresponding to the eigenvalue E_r. The mean value of the energy is given by

$$\langle H \rangle \equiv \overline{E} = \sum_r p_r E_r = -\frac{\partial}{\partial \beta} \ln Z. \tag{1.6}$$

In general, for any arbitrary function of the energy $f(E)$, its statistical average is given by

$$\overline{f(E)} = \sum_r p_r f(E_r). \tag{1.7}$$

The entropy of this system is given by

$$S = -k_B \sum_r p_r \ln p_r = -k_B \overline{\ln \rho}, \tag{1.8}$$

where ρ is the statistical density operator associated to the state under consideration. This is to be defined in Chap. 3. We shall see later that this logarithmic function of ρ is well defined from a rigorous mathematical setting.

After (1.8) and using (1.3), the entropy can be written as

$$S = -k_B \sum_r \frac{e^{-\beta E_r}}{Z} (-\beta E_r - \ln Z)$$

$$= k_B \beta \left(\sum_r E_r \frac{e^{-\beta E_r}}{Z} \right) + k_B \ln Z \sum_r \left(\frac{e^{-\beta E_r}}{Z} \right) = k_B \beta \overline{E} + k_B \ln Z, \tag{1.9}$$

so that

$$\left(\frac{S}{k_B \beta} \right) = \overline{E} + \frac{1}{\beta} \ln Z \iff \left(-\frac{1}{\beta} \ln Z \right) = \overline{E} - \frac{1}{k_B \beta} S. \tag{1.10}$$

Taken the derivative of the entropy S with respect to the energy $\overline{E}$, we obtain

$$\left(\frac{\partial S}{\partial \overline{E}} \right) = k_B \beta = \frac{1}{T} \implies k_B T = \frac{1}{\beta}. \tag{1.11}$$

After (1.10), we have

$$\overline{E} - TS = -(k_B T) \ln Z. \tag{1.12}$$

This expression is the well-known Helmholtz free energy :

$$\boxed{F := \overline{E} - TS = -(k_B T) \ln Z.} \tag{1.13}$$

The above expression has been obtained by assuming a definite structure for the entropy and the mean energy, adopting Boltzmann concept of probability. If, instead, we would like to determine the explicit expression for the probabilities, we should start from the following definition:

$$F = \sum_i (E_i\, p_i + k_B T\, p_i \ln p_i) \,, \tag{1.14}$$

and make the variation

$$\frac{\delta F}{\delta p_k} = \sum_j \left(E_j + \frac{1}{\beta} \ln p_j + \frac{1}{\beta}\right) \frac{\delta p_j}{\delta p_k} = 0 \,, \tag{1.15}$$

leading to

$$C_0 + \beta E_j + \ln p_j = 0 \,, \qquad \forall j \,, \tag{1.16}$$

where C_0 is a constant. Equivalently,

$$C_0 \times p_j = e^{-\beta E_j} \iff p_j = \frac{1}{C_0} e^{-\beta E_j} \,. \tag{1.17}$$

Since,

$$\sum_j p_j = 1 \iff C_0 = \sum_j e^{-\beta E_j} \,, \tag{1.18}$$

and

$$p_j = \frac{e^{-\beta E_j}}{\sum_k e^{-\beta E_k}} \,, \tag{1.19}$$

which is the form already adopted for the probability at equilibrium.

On the Mean Values and Standard Deviations

From the mean value of the energy, $\overline{E}$, and for a given value of the energy, E, its deviation is written as $\Delta E := E - \overline{E}$. Then,

$$(\Delta E)^2 = (E - \overline{E})^2 = E^2 + (\overline{E})^2 - 2E\overline{E} = (\overline{E^2}) - (\overline{E})^2 \tag{1.20}$$

is the standard deviation of the energy.

Taking into account the definition of Z in (1.4), and performing a second derivative with respect to β, one gets

$$\frac{\partial^2 \ln Z}{\partial \beta^2} = -\frac{1}{Z^2} \left(\sum_r E_r\, e^{-\beta E_r}\right)^2 + \frac{1}{Z} \sum_r E_r^2\, e^{-\beta E_r}$$

$$= -(\overline{E})^2 + (\overline{E^2}) = (\Delta E)^2 \,, \tag{1.21}$$

so that

$$(\Delta E)^2 = \frac{\partial^2 \ln Z}{\partial \beta^2} = -\frac{\partial \overline{E}}{\partial \beta} = -\left(\frac{\partial \overline{E}}{\partial T}\right) \frac{dT}{d\beta} = k_B T^2 C_V \,, \tag{1.22}$$

where we have defined the specific heat at constant volume as

$$C_V := \left(\frac{\partial \overline{E}}{\partial T}\right)_V . \tag{1.23}$$

With this definition for the specific heat, the *standard deviation* becomes

$$\left(\frac{\Delta E}{\overline{E}}\right) = \frac{(k_B T^2 C_V)^{1/2}}{\overline{E}} . \tag{1.24}$$

The variables C_V and $\overline{E}$ vary with the total number of particles, N, in the system. We shall denote this dependence as

$$C_V = \mathcal{O}(N) \quad \text{and} \quad \overline{E} = \mathcal{O}(N) , \tag{1.25}$$

where $\mathcal{O}$ represents "order of magnitude", following the notation of Landau. Then,

$$\left(\frac{\Delta E}{\overline{E}}\right) = \frac{(k_B T^2 C_V)^{1/2}}{\overline{E}} \approx \mathcal{O}(N^{1/2}/N) = \mathcal{O}(1/\sqrt{N}) , \tag{1.26}$$

so that the standard deviation of the energy decreases with the inverse of the square root of the number of particles.

An Example

Let us consider a system of N dipoles immersed in a constant magnetic field B. The dipoles are either parallel or antiparallel with respect to the direction of the field. Thus, the number of states accessible to each dipole is two: either parallel or antiparallel. The total number of microstates is equal to 2^N. The number $\Omega(n)$ of microstates with n parallel dipoles is the combinatory number

$$\Omega(n) = \binom{N}{n} = \frac{N!}{n!\,(N-n)!} . \tag{1.27}$$

The energy of the configuration of n dipoles parallel to the magnetic field is $E(n) = \mu B n$, and the energy of $N - n$ dipoles antiparallel is $E(N - n) = \mu B (N - n)$. The energy of the system is the sum of both terms: $E(n) + E(N - n)$, and this is written as

$$E(n, N) = n(-\mu B) + (N - n)(\mu B) = (N - 2n)\mu B , \tag{1.28}$$

where μ is the magnetic dipole moment of each particle. Next, from (1.27), we calculate the entropy of the same configuration

$$S(n, N) = k_B \ln \Omega(n) = k_B \ln N! - k_B \ln n! - k_B \ln(N - n)! . \tag{1.29}$$

If we apply in (1.29) the Stirling formula for the logarithm, $\ln q! \approx q \ln q - q$, then

$$S(n, N) \cong k_B\{N \ln N - N - n \ln n + n - (N - n) \ln(N - n) + (N - n)\}$$
$$\cong k_B\{N \ln N - n \ln n - (N - n) \ln(N - n)\}, \quad (1.30)$$

where the equal sign applies for large values of N. In real systems this approximation is valid, since N is of the order of the Avogadro number. To calculate the temperature of the system, we shall express (1.11) in terms of $\frac{dS(n,N)}{dn}$ and $\frac{dE(n,N)}{dn}$. The derivative of the entropy with respect to n at fixed N reads

$$\left(\frac{dS}{dn}\right)_N = k_B\{-\ln n - 1 + 1 + \ln(N - n)\} = k_B \ln\left(\frac{N - n}{n}\right), \quad (1.31)$$

and the derivative of the energy (1.28) is

$$\frac{dE(n, N)}{dn} = -2\varepsilon, \quad (1.32)$$

where $\varepsilon = \mu B$. In consequence,

$$\left(\frac{dS}{dE}\right)_{(n,N)} = \left[\left(\frac{dS}{dn}\right)\left(\frac{1}{\frac{dE}{dn}}\right)\right]_{(n,N)} = -\left(\frac{k_B}{2\varepsilon}\right) \ln\left(\frac{N - n}{n}\right)$$
$$= \left(\frac{k_B}{2\varepsilon}\right) \ln\left(\frac{n}{N - n}\right) = \frac{1}{T}. \quad (1.33)$$

Equation (1.33) implies

$$\boxed{\left(\frac{k_B T}{2\varepsilon}\right) \ln\left(\frac{n}{N - n}\right) = 1}. \quad (1.34)$$

After some straightforward manipulations, (1.34) can be cast in a more illustrative way, namely,

$$\frac{n}{N - n} = e^{2\varepsilon/k_B T} = \frac{n + N - N}{N - n} = \frac{N}{N - n} - 1 \Longrightarrow \frac{N}{N - n} = 1 + e^{2\varepsilon/kT}$$
$$\Longrightarrow \frac{N - n}{N} = 1 - \frac{n}{N} = \frac{1}{1 + e^{2\varepsilon/k_B T}} \Longrightarrow \frac{n}{N} = 1 - \frac{1}{1 + e^{2\varepsilon/k_B T}} = \frac{e^{2\varepsilon/k_B T}}{1 + e^{2\varepsilon/k_B T}}$$
$$= \frac{e^{\varepsilon/k_B T}}{e^{\varepsilon/k_B T} + e^{-\varepsilon/k_B T}} = \frac{1}{2}(1 + \tanh x)_{x=\varepsilon/k_B T}. \quad (1.35)$$

Clearly, for large values of $x = \varepsilon/k_B T$ the ratio n/N is close to one. For each dipole, the partition function reads

$$Z_1 = e^{\varepsilon/k_B T} + e^{-\varepsilon/k_B T} = e^x + e^{-x} = 2 \cosh x. \quad (1.36)$$

For $N = 2n$ distinguishable particles with no interaction among them, the partition function becomes

$$Z(N) = 2^N \cosh^N x = (Z_1)^N, \tag{1.37}$$

so that

$$\ln Z(N) = N \ln 2 + N \ln \cosh x. \tag{1.38}$$

In order to obtain the mean value of the energy $\overline{E}(N)$, we proceed as before and write

$$\overline{E}(N) = -\frac{\partial}{\partial \beta} \ln Z(N) = \left[-\varepsilon \frac{N \sinh x}{\cosh x} \right]_{x = \varepsilon\beta} = -(N\varepsilon) \tanh(\varepsilon/k_B T). \tag{1.39}$$

In the limit $x = \varepsilon/kT >> 1$, we have

$$\overline{E}(N) \approx -N\varepsilon \approx E(n = N), \tag{1.40}$$

which is the energy of the microstate with $n = N$ dipoles parallel oriented with respect to the magnetic field.

The variation of the mean energy with respect to the temperature T can now be easily calculated. The result is

$$C_V \equiv \frac{\partial \overline{E}}{\partial T} = (-N\varepsilon) \left[\frac{1}{\cosh^2(\varepsilon/k_B T)} \right] \left(\cosh^2(\varepsilon/k_B T) - \sinh^2(\varepsilon/k_B T) \right) \left(-\frac{\varepsilon}{k_B T^2} \right)$$

$$= N k_B \left(\frac{\varepsilon}{k_B T} \right)^2 \left[\frac{1}{\cosh^2(\varepsilon/k_B T)} \right] \tag{1.41}$$

We are now in the position to calculate the free energy:

$$F = -N k_B T \ln(2 \cosh(\varepsilon/k_B T)) \tag{1.42}$$

and the entropy:

$$S = \frac{1}{T}[E - F] = \frac{1}{T} \left[-N\varepsilon \tanh(\varepsilon/k_B T) + N k_B T \ln(2 \cosh(\varepsilon/k_B T)) \right]$$

$$= N k_B \left[\ln 2 + \ln \cosh(\varepsilon/k_B T)) - (\varepsilon/k_B T) \tanh(\varepsilon/k_B T) \right]. \tag{1.43}$$

The model, so far introduced, describes in a simple way the physics of paramagnetic systems. The statistical description of a paramagnetic solid follows from the set of equations:

Mean Energy Per Particle:

$$\left(\frac{\overline{E}}{N} \right) = -\varepsilon \tanh \left(\frac{\varepsilon}{k_B T} \right). \tag{1.44}$$

Specific heat at constant volume per particle:

$$\left(\frac{C_V}{Nk_B}\right) = \left(\frac{\varepsilon}{k_B T}\right)^2 \operatorname{sech}^2\left(\frac{\varepsilon}{k_B T}\right) . \tag{1.45}$$

Entropy per particle:

$$\left(\frac{S}{Nk_B}\right) = \ln 2 + \ln\cosh\left(\frac{\varepsilon}{k_B T}\right) - \left(\frac{\varepsilon}{k_B T}\right)\tanh\left(\frac{\varepsilon}{k_B T}\right) . \tag{1.46}$$

Free energy for particle:

$$\left(\frac{F}{N}\right) = -k_B T \ln\left(2\cosh\left(\frac{\varepsilon}{k_B T}\right)\right) . \tag{1.47}$$

The average dipole moment per particle is

$$\overline{\mu} = \frac{1}{\beta}\left(\frac{\partial \ln Z}{\partial B}\right) , \tag{1.48}$$

and the mean value for the system of N particles is

$$M := N\overline{\mu} = \frac{2Nk_B T}{2\cosh(\varepsilon/k_B T)}\sinh(\varepsilon/k_B T)\frac{\mu}{k_B T} = N\mu\tanh(\varepsilon/k_B T) , \tag{1.49}$$

so that

$$\frac{M}{N} = \mu\tanh\left(\frac{\varepsilon}{k_B T}\right) . \tag{1.50}$$

Starting with the definitions of entropy and energy and using (1.27), we can determine the regime of *thermodynamic* temperatures of the system:

$$\frac{1}{T} = \frac{\partial S}{\partial \overline{E}} = k_B \frac{\partial \ln \Omega}{\partial \overline{E}} . \tag{1.51}$$

In terms of the density of states we write

$$\frac{1}{k_B T} = \frac{1}{2\mu B}\ln\left(\frac{n}{N-n}\right) , \tag{1.52}$$

where, as before, n is the number of dipoles parallel to the magnetic field.

According to (1.52) and considering that $\mu > 0$ and $B > 0$, one sees that for $n \geq N/2$, the temperature T is positive, and for $n < N/2$, the temperature T becomes negative. Therefore, the notion of negative temperature associated to the population of states antiparallel to the magnetic field becomes meaningful. This result is more general since the regime of negative temperatures is present in systems which have an upper bound in the energy [4].

A Second Example

The second application of the basic statistical concepts is a system of non-interacting atoms oscillating around positions of equilibrium. For simplicity, we shall assume that these atoms are placed along a line so that the model is unidimensional. For each atom, the energy levels are those of the harmonic oscillator

$$\mathcal{E}_n = \hbar\omega \left(n + \frac{1}{2} \right), \qquad n = 0, 1, \dots, \tag{1.53}$$

where ω is the frequency and n is the quantum number. The corresponding partition function is written as

$$Z = \sum_{n=0}^{\infty} e^{-\beta \mathcal{E}_n} = e^{-\hbar\omega/2k_B T} \sum_{n=0}^{\infty} e^{-n\hbar\omega/k_B T} = e^{-\hbar\omega/2k_B T} \frac{1}{1 - e^{-\hbar\omega/k_B T}}$$

$$= \frac{e^{-x/2}}{e^{-x/2}(e^{x/2} - e^{-x/2})} = \frac{1}{2\sinh(x/2)} = \frac{1}{2}\,\text{cosech}\left(\frac{x}{2}\right), \tag{1.54}$$

where $x = \hbar\omega/k_B T$. For a single oscillator, the free energy is written as

$$F = k_B T \ln 2 + k_B T \ln\left(\sinh\left(\frac{\hbar\omega}{2k_B T} \right) \right). \tag{1.55}$$

Analogously, the average energy per oscillator is (recall that $\beta = 1/k_B T$)

$$\overline{F} = -\frac{\partial \ln Z}{\partial \beta} - \frac{\partial \ln Z}{\partial x}\left(\frac{\partial x}{\partial \beta} \right) = \left(\frac{\hbar\omega}{2} \right) \frac{1}{2\sinh\left(\frac{\hbar\omega\beta}{2} \right)} 2\cosh\left(\frac{\hbar\omega\beta}{2} \right)$$

$$= \frac{\hbar\omega}{2} \coth \frac{\hbar\omega}{2k_B T} = \frac{\hbar\omega}{2} + \frac{\hbar\omega}{e^{\beta\hbar\omega} - 1}. \tag{1.56}$$

The specific heat at constant volume for the three-dimensional case is given by

$$C_V = \frac{\partial \overline{E}}{\partial T} = 3N \left(\frac{\partial \mathcal{E}}{\partial \beta} \right)\left(\frac{\partial \beta}{\partial T} \right) = \frac{3Nk_B}{k_B^2 T^2} (\hbar\omega)^2 [e^{\beta\hbar\omega} - 1]^{-2} e^{\beta\hbar\omega}$$

$$= 3Nk \left(\frac{\hbar\omega}{k_B T} \right)^2 \frac{e^{\hbar\omega/k_B T}}{(e^{\hbar\omega/k_B T} - 1)^2}. \tag{1.57}$$

Note that the frequency ω depends on the oscillation parameter, b_0, which fixes the length scale of the oscillations ($b_0 = \sqrt{\frac{\hbar}{m\omega}}$). Let us explore the limits of the specific heat C_V, for high and low temperatures. In the first case ($\hbar\omega << k_B T$), we use the approximation

$$e^{\hbar\omega/k_B T} \approx 1 + \hbar\omega/k_B T \tag{1.58}$$

in (1.57), to show that

$$C_V \approx 3Nk_B \left(1 + \hbar\omega/k_B T\right) \approx 3Nk_B \, . \tag{1.59}$$

For low temperatures, $\hbar\omega \gg k_B T$ and the specific heat at constant volume is

$$C_V \approx 3Nk_B \left(\frac{\hbar\omega}{k_B T}\right)^2 e^{-\hbar\omega/k_B T} \, , \tag{1.60}$$

which goes to zero as $x = \hbar\omega/k_B T$ goes to infinity. Then, it is seen that the Dulong-Petit law is obeyed [5]

In summary,

- The partition function *a la Boltzmann* is given by $Z = \sum_r e^{-\beta\mathcal{E}_r}$, where the index r denotes sum over all possible microstates, $\mathcal{E}_r$.
- The probability *a la Boltzmann* associated to each possible configuration of microstates is $p_r = Z^{-1} e^{-\beta\mathcal{E}_r}$.
- The mean value $\overline{X}$ of an extensive variable X with respect to the associated Lagrange multiplier α is given by the derivative

$$\overline{X} = -\frac{\partial \ln Z}{\partial \alpha} \, . \tag{1.61}$$

- The Boltzmann entropy is $S = k \ln \Omega$, where Ω is the statistical weight.
- In the equilibrium, the thermodynamic magnitudes are given in terms of the partition function as

$$F = -k_B T \ln Z$$
$$\overline{E} = -\frac{\partial}{\partial \beta} \ln Z \tag{1.62}$$
$$S = k_B \left(1 - \beta \frac{\partial}{\partial \beta}\right) \ln Z \, .$$

- The quadratic deviation of the magnitude X with respect to the variable α is given by

$$\frac{\partial^2 \ln Z}{\partial \alpha^2} \, . \tag{1.63}$$

In the examples considered earlier, i.e., dipoles in a magnetic field (a typical case of a two-level system) and the non-interacting one-dimensional harmonic oscillators, finite size effects manifest in the boundedness of the quadratic deviations. For these cases, the specific heat at constant volume is finite and reaches a maximum (Schottky effect). The appearance of the Schottky effect is a common feature in systems with a finite number of degrees of freedom [6]. This effect manifests itself as a peak in the specific heat, which should not be taken as a signal for a phase transition. Another example of this sort is the statistical treatment of the nuclear spectrum, for which the excitation energy as a function of the temperature displays an S shape, the derivative of which, C_V, reaches a maximum.

The Classical Gas

This is a system of particles for which we make the following restrictions:

- The interaction energy between particles is very small compared with the kinetic energy. Particles interact so weakly that could be considered as free particles.
- The state of the gas is determined by means of labeling particles occupying accessible energy levels. The energy of a given configuration, $E = E(n_1, n_2, \ldots)$, depends on the number n_r of particles in the state r.
- The total number of particles remains constant: $\sum_r n_r = N$.
- Particles are indistinguishable.

In this case, the partition function is

$$Z(N) = \frac{1}{N!} \left[\sum_r e^{-\beta \mathcal{E}_r} \right]^N , \tag{1.64}$$

where the factorial $N!$ in the denominator is introduced in order to include the indistinguishability of the particles. For each level r, the energy $\mathcal{E}_r$ can be divided into the sum of two terms, one reflects the translation energy $\mathcal{E}_{\text{trans}}(r)$ and the other is the energy associated to the internal degrees of freedom $\mathcal{E}_{\text{int}}(r)$. Thus,

$$\mathcal{E} = \mathcal{E}_{\text{trans}}(r) + \mathcal{E}_{\text{int}}(r) . \tag{1.65}$$

For a single particle, the partition function is then

$$\xi = \sum_r e^{-\beta \mathcal{E}_r} = \sum_r e^{-\beta (\mathcal{E}_{\text{trans}}(r) + \mathcal{E}_{\text{trans}}(r))} = \sum_{r(\text{trans})} e^{-\beta \mathcal{E}_{\text{trans}}(r)} \sum_{r(\text{int})} e^{-\beta \mathcal{E}_{\text{int}}(r)} = \xi_{\text{trans}} \cdot \xi_{\text{int}} , \tag{1.66}$$

We note that the system is confined in a region of finite volume V, so that the translational factor ξ_{trans} has the form

$$\xi_{trans} = \frac{1}{h_0^3} \int d\mathbf{q} \int d\mathbf{p} \, e^{-\beta \mathbf{p}^2/2m} = \frac{1}{h_0^3} 4\pi V \int_0^\infty dp \, p^2 \, e^{-\beta p^2/2m} . \tag{1.67}$$

Here, h_0 is a normalization constant and $p := |\mathbf{p}|$ denotes the modulus of the particle momentum $\mathbf{p}$. The factor 4π comes from the integration over angles. This integral is extended to the positive semi-axis $[0, \infty)$, since the momentum is arbitrary. Then, (1.67) is equal to

$$\xi_{\text{trans}} = \left(\frac{4\pi V}{h_0^3}\right)(-2m)\frac{\partial}{\partial\beta}\left[\int_0^\infty dp\, e^{-(\beta/2m)p^2}\right]$$

$$= \left(\frac{4\pi V}{h_0^3}\right)(-2m)\frac{\partial}{\partial\beta}\left[\frac{1}{2}\sqrt{\frac{\pi\, 2m}{\beta}}\right] = V\left(\frac{2\pi m k_B T}{h_0^2}\right)^{3/2}. \tag{1.68}$$

Consequently, the partition function for the classical gas with N particles is given by

$$Z(N) = \frac{1}{N!}\, V^N \left(\frac{2\pi m k_B T}{h_0^2}\right)^{3N/2}[\xi_{\text{int}}]^N. \tag{1.69}$$

The intrinsic contribution of the internal degrees of freedom becomes relevant particularly in the case of composite systems like atoms, molecules, the atomic nucleus, etc., where degrees of freedom other than the translational one, e.g., rotational, vibrations, spin degree of freedom, etc., contributed. If, in a first approximation, we neglect internal effects, we obtain the following values for the thermodynamic functions:

((a) Free energy:

$$F = -k_B T \ln Z(N) = -k_B T\left(\frac{3}{2}N\ln\left\{\frac{2\pi m k_B T}{h_0^2}\,V^{2/3}\right\} - \ln N!\right)$$

$$= k_B T \ln N! - \frac{3}{2}k_B T\, N\ln\left(\frac{2\pi m k_B T}{h_0^2}\,V^{2/3}\right). \tag{1.70}$$

(b) Pressure:
The thermodynamic *equation of state* gives the pressure in terms of the variation of the free energy with respect to the volume at constant temperature, i.e.,

$$p = -\left(\frac{\partial F}{\partial V}\right) = \frac{N k_B T}{V} \implies pV = N k_B T. \tag{1.71}$$

(c) Energy:
The mean value of the energy is

$$\overline{E} = -\frac{\partial}{\partial\beta}\ln Z(N) = \frac{3N}{2}k_B T. \tag{1.72}$$

(d) Specific heat:
The specific heat at constant volume is

$$C_V = \left(\frac{\partial\overline{E}}{\partial T}\right) = \frac{3N}{2}k_B. \tag{1.73}$$

(e) Entropy:

The expression of the entropy is written as

$$S = \frac{1}{T}(\overline{E} - F) = k_B \left(\ln Z - \frac{\beta}{Z} \frac{\partial Z}{\partial \beta} \right) = k_B \left(\ln Z + \frac{3N}{2} \right). \quad (1.74)$$

Observe that $S \longmapsto \infty$ when $T \longmapsto 0$, in contradiction with Nernst's law that establishes that S should go to zero as $T \longmapsto 0$. The origin of this problem lies in the passage to the continuum $\sum_r \longmapsto \int dp$, which is not a good approximation for low temperatures because the contribution of zero momentum is not zero due to the internal degrees of freedom.

In summary,

- The method based on the statistical averages gives correctly the classical limit. However, some inconsistencies appear in the low-temperature regime, which are understood in the context of the quantum statistical mechanics, as we shall see next.
- In most cases, a passage to the continuum is required. Among the procedures to be used to perform such a passage are the box quantization and the $\delta(E)$-type discretization.

1.2 Partition Function for a Gas of Photons

Let us assume electromagnetic radiation confined in a *cavity* at constant temperature T. The radiation is in equilibrium with the cavity walls. The number of carriers of radiation (photons) inside the cavity does not remain constant. At some moment, there are $\{n_1, n_2, \ldots, n_k, \ldots\}$ photons with energies $\{\mathcal{E}_1, \mathcal{E}_2, \ldots, \mathcal{E}_k, \ldots\}$, respectively. The total energy is then

$$E(n_1, n_2, \ldots, n_k, \ldots) = n_1 \mathcal{E}_1 + n_2 \mathcal{E}_2 + \cdots + n_k \mathcal{E}_k, \ldots. \quad (1.75)$$

The partition function for such a system is given by

$$Z(T, V) = \sum_{n_1=0}^{\infty} \sum_{n_2=0}^{\infty} \cdots \sum_{n_k=0}^{\infty} \cdots e^{-\beta(n_1 \mathcal{E}_1 + n_2 \mathcal{E}_2 + \cdots + n_k \mathcal{E}_k + \cdots)}$$

$$= \prod_{k=1}^{\infty} \left\{ \sum_{n_k=0}^{\infty} e^{-\beta n_k \mathcal{E}_k} \right\} = \prod_{k=1}^{\infty} \left\{ \frac{1}{1 - e^{-\beta \mathcal{E}_k}} \right\}, \quad (1.76)$$

so that

$$\ln Z(T, V) = - \sum_{k=1}^{\infty} \ln(1 - e^{-\beta \mathcal{E}_k}). \quad (1.77)$$

After (1.77), the average number of photons, $\bar{n}_k$, in the energy level $\mathcal{E}_k$, is

$$\bar{n}_k = -\frac{1}{\beta}\frac{\partial}{\partial \mathcal{E}_k}(\ln Z(T,V)) = \left(\frac{1}{e^{\beta \mathcal{E}_k}-1}\right). \qquad (1.78)$$

The value $\bar{n}_k$ is often called the *occupation number* of the kth level.

1.2.1 The Blackbody Radiation

The most common application of all of the above is probably the celebrated *blackbody radiation*. As is well known, it was the study of the energy spectrum of this particular phenomenon which initiated the consideration of quantum mechanics (Planck 1900). In this case, the energy levels are labeled by a continuous label ω so that $\mathcal{E}_\omega = \hbar\omega$, $\hbar$ being the Planck constant divided by 2π. Since the variable ω is continuous, we have to replace the sums in (1.76) by integration with respect to ω, so that

$$\sum_k \longmapsto \int f(\omega)\,d\omega. \qquad (1.79)$$

For each frequency ω, the value of the function $f(\omega)$ in (1.79) is

$$f(\omega)\,d\omega = \frac{V}{\pi^2 c^3}\,\omega^2\,d\omega. \qquad (1.80)$$

In deriving the expression for $f(\omega)$, one has to account for the polarization degree of freedom, the volume in coordinates and momentum space, and the relationship between the momentum and the frequency. Collecting all these elements, one gets

$$f(\omega)\,d\omega = 2\frac{d^3\mathbf{q}\,d^3\mathbf{p}}{(2\pi\hbar)^3} = \frac{2\left(\hbar\dfrac{d\omega}{c}\right)\cdot\left(\dfrac{\hbar^2\omega^2}{c^2}\right)\cdot 4\pi V}{(2\pi\hbar)^3}, \qquad (1.81)$$

since $p = \hbar k = \hbar\omega/c$, which readily yields (1.80).

The number of photons with frequencies in the interval $\omega + d\omega$ is, after (1.78) and (1.81), given by

$$dN_\omega = \bar{n}_\omega f(\omega)\,d\omega = \left(\frac{1}{e^{\beta\hbar\omega}-1}\right)\left(\frac{V}{\pi^2 c^3}\right)\omega^2\,d\omega. \qquad (1.82)$$

If we assume that the energy distribution is uniform over the volume V, it results that the energy emitted due to the radiation in the interval of frequencies between ω and $\omega + d\omega$ is equal to

$$dE_\omega = \hbar\omega\,dN_\omega = \left(\frac{V\hbar}{\pi^2 c^3}\right)\left(\frac{\omega^3\,d\omega}{e^{\beta\hbar\omega}-1}\right) = u(\omega,T)\,d\omega, \qquad (1.83)$$

which is the celebrated *Planck radiation law*.

In order to determine the total energy emitted by the blackbody at the equilibrium temperature T, we have to integrate (1.83) over all range of frequencies:

$$U(T) := \int_0^\infty u(\omega, T)\, d\omega = \frac{V\hbar}{\pi^2 c^3} \int_0^\infty \frac{\omega^3\, d\omega}{e^{\beta\hbar\omega} - 1}. \qquad (1.84)$$

Now, using the notation, $x = \beta\hbar\omega$, it yields

$$U(T) = \frac{V\hbar}{\pi^2 c^3} \cdot \frac{1}{(\beta\hbar)^4} \int_0^\infty \frac{x^3\, dx}{e^x - 1} \iff \frac{U(T)}{V} = aT^4, \qquad (1.85)$$

where the emissivity factor a is

$$a = \frac{\hbar}{\pi^2 c^3} \cdot \frac{k_B^4}{\hbar^4} \int_0^\infty \frac{x^3\, dx}{e^x - 1} = \frac{k_B^4 \pi^2}{15\, c^3 \hbar^3}. \qquad (1.86)$$

Equation (1.85) is the Stefan-Boltzmann law, and the factor a in (1.86) is the related to blackbody emissivity, σ, by the expression

$$\sigma \approx \frac{c}{4}\, a = 5.67 \times 10^{-8}\, \frac{\text{J}}{\text{m}^2\ \text{seg}\ {}^o\text{K}^4}. \qquad (1.87)$$

It is interesting to obtain the values of $dE_\omega = \mu(\omega, T)\, d\omega$ for small and large values of the frequency ω as compared to the temperature T. For the smaller range of frequencies, $\hbar\omega << kT$, we have

$$dE_\omega = u(\omega, T)\, d\omega \longmapsto \frac{\omega^2}{\pi^2 c^3} kT\, d\omega, \qquad (1.88)$$

which is Rayleigh law (Rayleigh 1920) law.

For high frequencies, $\hbar\omega >> kT$, we have

$$dE_\omega = u(\omega, T)\, d\omega \longmapsto \frac{\hbar\omega^3}{\pi^2 c^3} e^{-\beta\hbar\omega}\, d\omega, \qquad (1.89)$$

which is Thompson law (Thompson 1920) law.

The matching between the power law dependence (Rayleigh) and the exponential regime (Thompson) suggests the existence of a maximum for some values of ω. This value can be obtained as a solution of a transcendental equation, since it is determined by the condition

$$\left(\frac{\partial u(\omega, T)}{\partial \omega} \right)_T = 0. \qquad (1.90)$$

This gives

$$\frac{1}{(e^{\beta\hbar\omega} - 1)^2} \left\{ \left(e^{\beta\hbar\omega} - 1 \right) 3\omega^2 - \omega^3 e^{\beta\hbar\omega} \beta\hbar \right\} = 0, \qquad (1.91)$$

which yields to

$$e^{\beta\hbar\omega}(3 - \beta\hbar\omega) = 3 . \tag{1.92}$$

This transcendental equation has a solution that can be found as the intersection between two curves. The expression for $\beta\hbar\omega$ at the peak of the distribution is given by the Wien spectral displacement law :

$$\frac{\beta_1\hbar\omega_1}{\beta_2\hbar\omega_2} = 1 \iff \frac{\omega_1}{\omega_2} = \frac{T_1}{T_2} . \tag{1.93}$$

The logarithm of the partition function was given in (1.79). After the replacement of the discrete sum by the integral in the process of passage to the continuum and taking into account (1.80), we obtain

$$\ln Z(T, V) = -\sum_{k=1}^{\infty} \ln(1 - e^{-\beta\mathcal{E}_k}) = -\int_0^{\infty} \frac{V\omega^3}{\pi^2 c^3}(1 - e^{-\beta\hbar\omega})\,d\omega . \tag{1.94}$$

After the change of variable $x := \beta\hbar\omega$, (1.94) gives

$$\ln Z = \frac{V\omega^3}{\pi^2 c^3 (\beta\hbar)^3} \int_0^{\infty} x^2 \ln(1 - e^{-x})\,dx , \tag{1.95}$$

so that the free energy $F = -k_B T \ln Z$ has the following expression:

$$F = VT^4 \left(\frac{k_B^4}{\pi^2 c^3 \hbar^3}\right) \int_0^{\infty} x^2 \ln(1 - e^{-x})\,dx . \tag{1.96}$$

The integral in (1.96) can be easily solved by parts. If we call this integral I, we have

$$I = \frac{x^3}{3} \ln(1 - e^{-x})\Big|_0^{\infty} - \frac{1}{3} \int_0^{\infty} \frac{x^3 e^{-x}}{1 - e^{-x}}\,dx . \tag{1.97}$$

The first expression in (1.97) vanishes. The limit at $+\infty$ can be calculated using the l'Hopital rule. If we divide by e^{-x} the quotient under the integral sign in (1.97), we find that

$$I = -\frac{1}{3} \int_0^{\infty} \frac{x^3}{e^x - 1}\,dx = -\frac{1}{3} \times \frac{\pi^4}{15} . \tag{1.98}$$

Therefore,

$$F = VT^4 \frac{k_B^4}{\pi^2 c^3 \hbar^3} \left(-\frac{1}{3} \times \frac{\pi^4}{15}\right) = -\left(\frac{k_B^4 \pi^2}{15 c^3 \hbar^3}\right) \frac{1}{3} VT^4 = -\frac{1}{3} aVT^4 , \tag{1.99}$$

where the definition of a in (1.103) is clear. From (1.99), we obtain the following equations:

$$S = -\left(\frac{\partial F}{\partial T}\right)_V = \frac{4}{3} aVT^3,\qquad(1.100)$$

$$P = -\left(\frac{\partial F}{\partial V}\right)_T = \frac{1}{3} aT^4.\qquad(1.101)$$

Therefore, the average of the energy is given by

$$\overline{E} = F + TS = -\frac{a}{3} VT^4 + \frac{4}{3} aVT^4 = aVT^4,\qquad(1.102)$$

so that the pressure becomes

$$p = \frac{1}{3}\frac{\overline{E}}{V}.\qquad(1.103)$$

The expression (1.83) known as the Planck distribution law is an extension of the Bose–Einstein distribution law as we shall see in the forthcoming sections. Generally speaking, the Planck distribution law describes massless bosons.

1.3 The Maxwell Distribution

In the present section, we consider a gas formed by N identical particles in thermal equilibrium at temperature T. The average number of particles in a state with momentum $\mathbf{p}$ is given by $\overline{n}(\mathbf{p})$. The kinetic energy of a three-dimensional particle with momentum $\mathbf{p}$ is $\mathcal{E}(\mathbf{p}) = \mathbf{p}^2/2m$, so that the normalized probability distribution is ($p := |\mathbf{p}|$)

$$C \int_0^\infty 4\pi\, p^2 e^{-p^2/(2mkT)}\, dp = 1,\qquad(1.104)$$

where C is a normalization constant. After the change of variables given by

$$x := \left(\frac{1}{2mk_BT}\right) p,\qquad(1.105)$$

(1.105) becomes

$$1 = 4\pi C (2mk_BT)^{3/2} \int_0^\infty x^2 e^{-x^2}\, dx = 4\pi C (2mk_BT)^{3/2} \left(-\frac{\partial}{\partial\alpha}\left[\int_0^\infty e^{-\alpha x^2}\, dx\right]\right)$$

$$= 4\pi C (2mk_BT)^{3/2}\left(-\frac{\partial}{\partial\alpha}\left[\frac{1}{2}\sqrt{\frac{\pi}{\alpha}}\right]\right) = C\,(2m\pi k_BT)^{3/2},\qquad(1.106)$$

for $\alpha = 1$.

Consequently,

$$C = (2m\pi k_B T)^{-3/2}.$$ (1.107)

Therefore, the probability of finding particles with the modulus of the momentum between p and $p + dp$ is

$$P(p)\,dp = \frac{4\pi\,p^2\,e^{-p^2/2mk_B T}\,dp}{(2m\pi k_B T)^{3/2}}.$$ (1.108)

Since all particles have mass m, then $p = mv$, where v is the modulus of the velocity of the particle with momentum **p**. The probability (1.108) in terms of the variable v is

$$P(v)\,dv = 4\pi v^2 \left(\frac{m}{2\pi k_B T}\right)^{3/2} e^{-mv^2/2kT}\,dv.$$ (1.109)

If we define the variable μ as

$$\mu := \left(\frac{m}{2k_B T}\right)^{1/2} v,$$ (1.110)

the expression (1.109) becomes

$$P(\mu)\,d\mu = \frac{4}{\sqrt{\pi}}\,\mu^2\,e^{-\mu^2}\,d\mu.$$ (1.111)

It is interesting to obtain the values of the maximum μ_M, the average $\overline{\mu}$ and the quadratic average $\overline{\mu^2}$ corresponding for $P(\mu)$:

Maximum (μ_M):
The maximum of $P(\mu)$ is obtained as a solution of the equation

$$\frac{\partial P(\mu)}{\partial \mu} = 0.$$ (1.112)

Thus, we have to solve

$$\frac{\partial P(\mu)}{\partial \mu} = \frac{4}{\sqrt{\pi}}\left[2\mu e^{-\mu^2} + \mu^2(-2\mu e^{-\mu^2})\right] = \frac{4}{\sqrt{\pi}}\left[2\mu e^{-\mu^2}(1-\mu^2)\right] = 0.$$ (1.113)

Since μ must be positive, this maximum value is reached at one, which also gives the maximal speed v_M

$$\mu_M = 1 \implies v_M = \left(\frac{2k_B T}{m}\right)^{1/2}.$$ (1.114)

Average $(\overline{\mu})$

The average or mean value of μ with distribution function $P(\mu)$ is easily calculated:

$$\overline{\mu} = \frac{4}{\sqrt{\pi}} \int_0^\infty \mu\mu^2 e^{-\mu^2}\, d\mu = \frac{4}{\sqrt{\pi}}\frac{1}{2} = \frac{2}{\sqrt{\pi}}\,. \tag{1.115}$$

Then, using (1.110) and (1.115), we obtain the mean value of the particle speed in terms of the maximal speed v_M as

$$\overline{v} = \left(\frac{2k_B T}{m}\right)^{1/2}\frac{2}{\sqrt{\pi}} = \frac{2}{\sqrt{\pi}}\, v_M\,. \tag{1.116}$$

Quadratic Average $(\overline{\mu^2})$

This is obtained as

$$\overline{\mu^2} = \frac{4}{\sqrt{\pi}} \int_0^\infty \mu^4 e^{-\mu^2}\, d\mu = \frac{3}{2}\,, \tag{1.117}$$

so that

$$\overline{v^2} = \frac{3}{2}\, v_M^2\,. \tag{1.118}$$

Now, from (1.104) and (1.107), we write the distribution function in terms of the energy $E = p^2/2m$, which is

$$P(E)\, dE = \frac{2}{\sqrt{\pi}} \frac{E^{1/2} e^{-E/k_B T}}{(k_D T)^{3/2}}\, dE\,. \tag{1.119}$$

In the sequel, we shall often use the notation $\mathcal{E} = E/k_B T$, so that the distribution $P(E)$ takes the form

$$P(\mathcal{E}) = \frac{2}{\sqrt{\pi}}\, \mathcal{E}^{1/2} e^{-\mathcal{E}}\,. \tag{1.120}$$

Let us calculate the mean value of the energy for a system which obeys the distribution function $P(E)$:

$$\overline{E} = \int_0^\infty E\, P(E)\, dE = \frac{2}{\sqrt{\pi}} \int_0^\infty \frac{E^{3/2}}{(k_B T)^{3/2}} e^{-E/k_B T}\, dE$$
$$= \frac{2}{\sqrt{\pi}}\, (k_B T) \int_0^\infty \mathcal{E}^{3/2} e^{-\mathcal{E}}\, d\mathcal{E}\,. \tag{1.121}$$

In order to obtain the value of the last integral, let us recall the definition of the *Gamma function*, $\Gamma(x)$, that for real non-negative values of x, i.e., $x > 0$, takes the following form:

$$\Gamma(x) = \int_0^\infty t^{x-1} e^{-t}\, dt\,. \tag{1.122}$$

The function $\Gamma(x)$ has the following properties:

$$\Gamma(x+1) = x\,\Gamma(x)\,, \quad x > 0\,, \quad \Gamma\left(\frac{1}{2}\right) = \sqrt{\pi}\,. \tag{1.123}$$

Note that formula (1.122) along (1.121) imply that

$$\overline{E} = \frac{2}{\sqrt{\pi}}\,(k_B T)\Gamma\left(\frac{5}{2}\right) = \frac{3}{2}\,k_B T\,, \tag{1.124}$$

where the last identity in (1.124) has been obtained using (1.123). This result agrees with that obtained using the partition function method in the classical gas.

For the quadratic average, we get

$$\overline{E^2} = \int_0^\infty E^2\,P(E)\,dE = \frac{2}{\sqrt{\pi}}\,(k_B T)^2 \int_0^\infty \mathcal{E}^{5/2}\,e^{-\mathcal{E}}\,d\mathcal{E}$$

$$= \frac{2}{\sqrt{\pi}}\,(k_B T)^2\Gamma\left(\frac{7}{2}\right) = \frac{15}{4}(k_B T)^2\,. \tag{1.125}$$

Recall that the mean square deviation is given by $(\Delta E)^2 = \overline{E^2} - (\overline{E})^2$. For a system of N particles, we have to multiply this deviation by N, so that (1.124) and (1.125) yield

$$\frac{\Delta E}{E} = \sqrt{\frac{2}{3}}\,\frac{1}{\sqrt{N}}\,. \tag{1.126}$$

Then, the energy fluctuations are of the order $O(N^{-1/2})$.

1.4   Classical Statistical Mechanics

From the previously discussed examples, it becomes clear that the partition function

$$Z = \sum_{\{E\}} g(E)\,e^{-\beta E}\,, \tag{1.127}$$

where $\{E\}$ stands for the energy of all accessible states, may be used to obtain the *classical* results of Maxwell as well as the (pre-quantum) Boltzmann and Gibbs formulations [7]. From the classical point of view, this sum is given by

$$Z = \frac{1}{h_0^\nu} \int dp_1\ldots,dp_\nu\,dq_1\ldots dq_\nu\,\,e^{-\beta H(p_1,\ldots,p_\nu,q_1,\ldots,q_\nu)}\,, \tag{1.128}$$

which corresponds to a classical system with ν degrees of freedom. The constant h_0 is the unit volume in phase space.

1.4.1 On the Equipartition Theorem

Let us consider a system in thermal equilibrium at temperature T. Assume that the Hamiltonian depends quadratically on the coordinates and momenta. To find out the contribution of a given coordinate, say q_1, to the total energy of the system we write

$$H = A\, q_1^2 + H', \tag{1.129}$$

where A is a constant and H' depends on the momenta and the remaining coordinates. Therefore,

$$e^{-\beta H} = e^{-\beta H'}\, e^{-\beta A q_1^2}. \tag{1.130}$$

Due to the separability, the partition function is a product of an integral in which the variable q_1 has been excluded,

$$Z' = \int dq_2 \dots dq_\nu\, dp_1 \dots dp_\nu\, e^{-\beta H'} \tag{1.131}$$

and a term in q_1:

$$Z_1 := \int_{-\infty}^{\infty} e^{-\beta A q_1^2}\, dq_1. \tag{1.132}$$

The corresponding normalized distribution function is

$$P(q_1) = [Z_1^{-1}\, e^{-\beta A q_1^2}]\, [Z'^{-1}\, e^{-\beta H'}]. \tag{1.133}$$

Then, we can evaluate the contribution of the term $A\, q_1^2$ in H to the energy average as

$$\int (A q_1^2)\, P(q_1)\, dq_1\, dq_2 \dots dq_\nu\, dp_1 \dots dp_\nu$$

$$= \left[\frac{A}{Z_1} \int q_1^2\, e^{-\beta A q_1^2}\, dq_1 \right] \left[\frac{1}{Z'} \int e^{-\beta H'}\, dq_2 \dots dq_\nu\, dp_1 \dots dp_\nu \right]$$

$$= \left[\frac{A}{Z_1} \int_{-\infty}^{\infty} q_1^2\, e^{-\beta A q_1^2}\, dq_1 \right] \frac{Z'}{Z'} = -\frac{\partial}{\partial \beta} \ln \left[\int_{-\infty}^{\infty} e^{-\beta A q_1^2}\, dq_1 \right]$$

$$= -\frac{\partial}{\partial \beta} \ln \left[\sqrt{\frac{\pi}{A\beta}} \right] = \frac{1}{2\beta} = \frac{k_B T}{2}. \tag{1.134}$$

In deriving the above equation, it is necessary to assume that the separation of the coordinate (or the momentum) q_i (p_i) does not affect the integration over the remainder variables.

To illustrate the importance of this result, we consider the case of the harmonic oscillator in $6N$ variables.

The Hamiltonian which describes N three-dimensional uncoupled harmonic oscillators of a common frequency ω has $3N$ coordinates and $3N$ conjugate momenta:

$$H = \sum_{i=1}^{N} \left(\frac{1}{2m} \mathbf{p}_i^2 + \frac{1}{2} m\omega^2 \mathbf{q}_i^2 \right) . \tag{1.135}$$

For it,

1. All coordinates and momenta are separated in the sense defined above.
2. All coordinates and momenta are quadratic.

Then, Eq. (1.134) applies for each of the coordinates and momenta. In consequence, the energy average for this system is

$$\overline{E} = 3N \left[\frac{1}{2} k_B T + \frac{1}{2} k_B T \right] = 3N k_B T . \tag{1.136}$$

This is the so-called *equipartition principle*.

So far, we have considered configuration *averages* through the partition function. In spite of the difficulties found in counting the exact number of degrees of freedom or the available energy states of a given system, the use of averages over replicas has proven to be much easier. The procedure amounts to the definition of average values and deviations as dictated by the rule so far considered. Now, the question concerns the validity of the procedure with the use of operators instead of variables. This will be discussed in the next section.

Concerning the quantum formulation of statistical mechanics, we should consider three different aspects: (i) densities of states, (ii) time and thermal evolution of operators, and (iii) symmetry properties of states.

In the sequel, we intend to develop a formulation of quantum statistical mechanics based on these concepts and we shall recover the classical results as classical limits of quantum ones.

1.5 Systems with a Variable Number of Particles

The partition function depends on the energy and the number of particles. Linked to these extensive magnitudes we have two intensive variables: temperature ($\beta = 1/k_B T$) and chemical potential μ, respectively. In this situation, the partition function should have the form

$$Z(\beta, \lambda) = \sum_{\{E,N\}} e^{-\beta E + \lambda N} \tag{1.137}$$

where the sum extends to all energy configurations and all possible distributions of particles among the energy levels.

Let us assume first that the number of particles has been fixed to be N. These particles are distributed in different energy levels: n_1 particles with energy $\mathcal{E}_1$, n_2 with energy $\mathcal{E}_2$ and so on, so that $N = \sum_r n_r$. If the particles are not interacting, it is reasonable to assume that the sum of the energies of each level is equal to the total energy, i.e.,

$$E = \sum_r n_r \,\mathcal{E}_r \,, \qquad N = \sum_r n_r \,, \qquad E - \mu N = \sum_r \left[n_r \mathcal{E}_r - \mu n_r \right] \,, \qquad (1.138)$$

where the index r labels the energy levels. Then, the partition function becomes

$$Z(\beta, \mu) = \sum_{\{n_r\}} e^{-\sum_r \beta(\mathcal{E}_r - \mu)n_r} = \sum_{\{n_r\}} \prod_r \left[e^{-\beta(\mathcal{E}_r - \mu)n_r} \right] \,, \qquad (1.139)$$

where $\{n_r\}$ represents the sum over all possible configurations a, and n_r is the occupation number of the level with energy $\mathcal{E}_r$.

In the quantum case, we shall deal with two kinds of identical particles, fermions and bosons. For fermions, energy levels are either empty or occupied by only one particle at the time, so that $n_r = 0, 1$. This restriction is the celebrated *Pauli exclusion principle*, and it does not apply for bosons, for which $n_r = 0, 1, 2, \ldots, N$. We shall assume that our particles are either fermions or bosons.

Assume that $r = 1, 2, \ldots, p$, so that the energies of our levels are $\mathcal{E}_1, \mathcal{E}_2, \ldots, \mathcal{E}_p$. When the particles are fermions, all these levels may be occupied either by no particle or just by one particle. The allowed configurations are depicted in the following table:

$$\left.\begin{matrix} 1 & 0 & 0 & \ldots & 0 \\ 0 & 1 & 0 & \ldots & 0 \\ \multicolumn{5}{c}{\cdots\cdots\cdots\cdots} \\ 0 & 0 & 0 & \ldots & 1 \end{matrix}\right\} N = 1$$

$$\left.\begin{matrix} 1 & 1 & 0 & \ldots & 0 \\ 1 & 0 & 1 & \ldots & 0 \\ \multicolumn{5}{c}{\cdots\cdots\cdots\cdots} \\ 0 & 0 & \ldots & 1 & 1 \end{matrix}\right\} N = 2$$

$$\cdots\cdots\cdots\cdots$$

$$\left.\begin{matrix} 1 & 1 & 1 & \ldots & 0 & 0 & 0 \\ \multicolumn{7}{c}{\cdots\cdots\cdots\cdots\cdots\cdots} \\ 0 & 0 & 0 & \ldots & 1 & 1 & 1 \end{matrix}\right\} N \,, \qquad (1.140)$$

with always $p \geq N$. The values of the first column represent the occupation numbers of the energy level $\mathcal{E}_1$, the values of the second one the occupational values of $\mathcal{E}_2$ and so on. The first N rows represent the configurations with one particle, the next $\binom{N}{2}$ are the configurations of 2 particles and so on until the last row that shows the unique configuration of N particles. Consequently, the partition function (1.139) for each array is written as

$$e^{-\beta(\mathcal{E}_1-\mu)1} e^{-\beta(\mathcal{E}_2-\mu)0} \ldots e^{-\beta(\mathcal{E}_p-\mu)0}$$

$$\ldots \quad \ldots \quad \ldots$$

$$e^{-\beta(\mathcal{E}_1-\mu)1} e^{-\beta(\mathcal{E}_2-\mu)1} \ldots e^{-\beta(\mathcal{E}_p-\mu)1}. \tag{1.141}$$

This product of exponential factors is equal to

$$Z(\beta, \mu) = \prod_r \left[1 + e^{-\beta(\mathcal{E}_r-\mu)} \right], \tag{1.142}$$

which is the expression for the partition function in the case of fermions.

In the case of bosons, we may depict a table like (1.130), where $0 \leq n_r \leq N$. The partition function becomes

$$Z(\beta, \mu) = \sum_{\{n_r\}} e^{-\beta(\mathcal{E}_1-\mu)n_1} e^{-\beta(\mathcal{E}_2-\mu)n_2} \ldots e^{-\beta(\mathcal{E}_p-\mu)n_p}, \tag{1.143}$$

which is nothing else that the product

$$Z(\beta, \mu) = \prod_r \left\{ \sum_{n_r=0}^{\infty} e^{-\beta(\mathcal{E}_r-\mu)n_r} \right\} = \prod_r \left[\frac{1}{1 - e^{-\beta(\mathcal{E}_r-\mu)}} \right]. \tag{1.144}$$

To find the average occupation numbers, for both bosons and fermions, we take logarithms in (1.142) and (1.144) so as to obtain

$$\ln Z(\beta, \mu) = \pm \sum_r \ln(1 \pm e^{-\beta(\mathcal{E}_r-\mu)}) \left\{ \begin{matrix} f \\ b \end{matrix} \right., \tag{1.145}$$

where f stands for fermions (plus signs) and b for bosons (minus signs).

Taking derivatives in (1.145), we obtain

$$-\frac{1}{\beta} \left(\frac{\partial \ln Z}{\partial \mathcal{E}_r} \right) = \sum_r \frac{e^{-\beta(\mathcal{E}_r-\mu)}}{1 \pm e^{-\beta(\mathcal{E}_r-\mu)}} = \sum_r \frac{1}{e^{\beta(\mathcal{E}_r-\mu)} \pm 1} \left\{ \begin{matrix} f \\ b \end{matrix} \right., \tag{1.146}$$

The average occupation numbers are

$$\overline{n}_r = \frac{1}{e^{\beta(\mathcal{E}_r-\mu)} + 1} \tag{1.147}$$

for fermions and

$$\bar{n}_r = \frac{1}{e^{\beta(\mathcal{E}_r - \mu)} - 1} \tag{1.148}$$

for bosons, respectively. It is important to remark that these results have been obtained under the assumption of particle *indistinguishability*. This is a general feature of quantum mechanics. However, in classical mechanics one may admit the possibility to distinguish particles, this happens in the so-called Maxwell–Boltzmann statistics.

1.5.1 On the Classical Limits of Quantum Statistics

In the classical case, we imagine a number of replicas of the system under study and make averages subject to various constraints such as number of particles, etc. Optimization with constraints requires one Lagrange multiplier for each constraint. In the case of fixed number of particles, the Lagrange multiplier is just $\alpha := -\beta\mu$. We write

$$\mathcal{Z}(\beta, \alpha) := \sum_{N=0}^{\infty} e^{-\alpha N} Z(N). \tag{1.149}$$

Note that the quantity $\mathcal{Z}(\beta, \alpha)$ is finite since it is a product of a fast-growing quantity, $Z(N)$, and a fast-decaying exponential.

Then, proceeding as in the previous sections, we shall specify the contributions to (1.149) considering the allowed values of the occupation numbers:

Maxwell–Boltzmann. Classical statistics with particle distinguishability and fixed number of particles equal to N. Then, $n_r = 0, 1, 2, \ldots, N$.

Planck. Quantum statistics, $n_r = 0, 1, 2, \ldots, \infty$, massless indistinguishable particles.

Bose–Einstein. Quantum statistics, $n_r = 0, 1, 2, \ldots, N$, massive indistinguishable particles.

Fermi–Dirac. Quantum statistics which obey the Pauli principle, $n_r = 0, 1$, either massive or massless indistinguishable particles.

Definition. The function $\mathcal{Z}(\beta, \alpha)$ in (1.137) is called the *grand partition function*.

Contrarily to the partition function which applies to systems with a fixed number of particles, the grand partition function refers to systems with arbitrary number of particles.

We have obtained the averages in the occupation numbers for the (quantum) Fermi–Dirac and Bose–Einstein statistics. It is interesting to obtain these averages for Maxwell–Boltzmann and Planck statistics. In the first case, the partition function corresponding to a fixed number, N, of particles is given by

$$Z(N) = \sum_{n_1, \ldots, n_m} \left[\frac{N!}{n_1! n_2! \ldots n_m!} \right] e^{-\beta\mathcal{E}_1 n_1 - \beta\mathcal{E}_2 n_2 - \cdots - \beta\mathcal{E}_m n_m} = \left(\sum_{i=1}^{m} e^{-\beta\mathcal{E}_i} \right)^N. \tag{1.150}$$

The factor depending on the number of particles in (1.150) accounts for the distinguishability of these particles. Expression (1.154) is often replaced by its corrected version $Z_C(N)$:

$$Z_C(N) = \frac{1}{N!} \left(\sum_{i=1}^{m} e^{-\beta \mathcal{E}_i} \right)^N , \tag{1.151}$$

and the grand partition function becomes

$$\mathcal{Z}(\beta, \alpha) = \sum_{N=0}^{\infty} \frac{1}{N!} \left(e^{-\alpha} \sum_{i=1}^{m} e^{-\beta \mathcal{E}_i} \right)^N = \sum_{N=0}^{\infty} \frac{1}{N!} \left(\sum_{i=1}^{m} e^{-\beta (\mathcal{E}_i - \mu)} \right)^N . \tag{1.152}$$

We compute the average occupational numbers as

$$\bar{n}_i = -\frac{1}{\beta} \frac{\partial \ln Z_C(N)}{\partial \mathcal{E}_i} = \frac{N e^{-\beta \mathcal{E}_i}}{\sum_i e^{-\beta \mathcal{E}_i}} . \tag{1.153}$$

For the Planck statistics the partition function is written as

$$Z = \sum_{n_1, \dots, n_\infty} e^{-\beta (\mathcal{E}_1 n_1 + \mathcal{E}_2 n_2 + \dots)}$$

$$= \sum_{n_1=0}^{\infty} e^{-\beta \mathcal{E}_1 n_1} \sum_{n_2=0}^{\infty} e^{-\beta \mathcal{E}_2 n_2} \dots \sum_{n_j=0}^{\infty} e^{-\beta \mathcal{E}_j n_j} \dots = \prod_{j=1}^{\infty} \frac{1}{1 - e^{-\beta \mathcal{E}_j}} , \tag{1.154}$$

so that

$$\ln Z = - \sum_{k=1}^{\infty} \ln(1 - e^{-\beta \mathcal{E}_k}) \tag{1.155}$$

and therefore, for all values of k, we have

$$\bar{n}_k = -\frac{1}{\beta} \frac{\partial \ln Z}{\partial \mathcal{E}_k} = \frac{1}{e^{\beta \mathcal{E}_k} - 1} , \tag{1.156}$$

which is the desired result.

1.5.1.1 Particle Number Dispersion

By definition, the dispersion in the particle number is given by the formula

$$\frac{\overline{(\Delta \bar{n}_i)^2}}{(\bar{n}_i)^2} , \tag{1.157}$$

with

$$\overline{(\Delta \bar{n}_i)^2} = -\frac{1}{\beta} \left(\frac{\partial \bar{n}_i}{\partial \mathcal{E}_i} \right) . \tag{1.158}$$

The dispersions (1.157) for the different statistics are shown next.

1.5.1.2 Maxwell–Boltzmann Statistics

The average $\bar{n}_i$ has been obtained in (1.153). Thus, using (1.158) and simplifying, we have

$$\overline{(\Delta \bar{n}_i)^2} = \bar{n}_i \left[1 - \frac{\bar{n}_i}{N} \right], \tag{1.159}$$

so that

$$\frac{\overline{(\Delta \bar{n}_i)^2}}{(\bar{n}_i)^2} = \frac{1}{\bar{n}_i} \left(1 - \frac{\bar{n}_i}{N} \right) = \frac{1}{\bar{n}_i} - \frac{1}{N}. \tag{1.160}$$

1.5.1.3 Planck Statistics

We have to use now (1.156) along with (1.157) and (1.160), so as to obtain

$$\frac{\overline{(\Delta \bar{n}_i)^2}}{(\bar{n}_i)^2} = 1 + \frac{1}{\bar{n}_i}. \tag{1.161}$$

1.5.1.4 Bose–Einstein Statistics

Take the expression (1.148) for $\bar{n}_i$. Note that μ should depend on $\mathcal{E}_i$ for all values of the index i. Therefore, (1.158) gives

$$\overline{(\Delta \bar{n}_i)^2} = \frac{e^{\beta(\mathcal{E}_i - \mu)}}{(e^{\beta(\mathcal{E}_i - \mu)} - 1)^2} \left\{ 1 - \frac{\partial \mu}{\partial \mathcal{E}_i} \right\}. \tag{1.162}$$

Since the number of particles N is fixed, we have to fulfil the condition

$$\frac{\partial N}{\partial \mathcal{E}_i} = 0. \tag{1.163}$$

Thus,

$$\frac{\partial N}{\partial \mathcal{E}_i} = -\beta \frac{e^{\beta(\mathcal{E}_i - \mu)}}{(e^{\beta(\mathcal{E}_i - \mu)} - 1)^2} - \beta \sum_j \frac{e^{\beta(\mathcal{E}_j - \mu)}}{(e^{\beta(\mathcal{E}_j - \mu)} - 1)^2} \left(-\frac{\partial \mu}{\partial \mathcal{E}_i} \right). \tag{1.164}$$

Since

$$\frac{e^{\beta(\mathcal{E}_i - \mu)}}{(e^{\beta(\mathcal{E}_i - \mu)} - 1)^2} = \frac{e^{\beta(\mathcal{E}_i - \mu)} - 1}{(e^{\beta(\mathcal{E}_i - \mu)} - 1)^2} + \frac{1}{(e^{\beta(\mathcal{E}_i - \mu)} - 1)^2} = \bar{n}_i + \bar{n}_i^2, \tag{1.165}$$

Equation (1.165) becomes

$$\bar{n}_i + \bar{n}_i^2 - \left(\sum_j (\bar{n}_i + \bar{n}_i^2) \frac{\partial \mu}{\partial \mathcal{E}_i} \right) = 0, \tag{1.166}$$

which gives

$$\frac{\partial \mu}{\partial \mathcal{E}_i} = \frac{\bar{n}_i(1+\bar{n}_i)}{\sum_j \bar{n}_j(1+\bar{n}_j)} . \tag{1.167}$$

By using (1.167) in (1.162) and dividing by the square of $\bar{n}_i$ so as to obtain

$$\frac{\overline{(\Delta\bar{n}_i)^2}}{(\bar{n}_i)^2} = \left(1+\frac{1}{\bar{n}_i}\right)\left[1-\frac{\bar{n}_i(1+\bar{n}_i)}{\sum_j \bar{n}_j(1+\bar{n}_j)}\right] . \tag{1.168}$$

1.5.1.5 Fermi–Dirac Statistics

Now, we use (1.148) for $\bar{n}_i$. Then, by a similar reasoning as in the previous case, we obtain

$$\frac{\overline{(\Delta\bar{n}_i)^2}}{(\bar{n}_i)^2} = \left(\frac{1}{\bar{n}_i}-1\right)\left[1-\frac{\bar{n}_i(1-\bar{n}_i)}{\sum_j \bar{n}_j(1+\bar{n}_j)}\right] . \tag{1.169}$$

The above results show that the dispersion (1.157) is minimal for the Fermi–Dirac statistics, for which $\bar{n}_i = 1$ for occupied levels and 0 otherwise.

On the other hand, we see that the deviations for Planck and Bose–Einstein statistics are bigger than those calculated with the Maxwell–Boltzmann statistics.

1.5.1.6 Quantum Statistics in the Classical Limit

We are in the position to analyze the classical limit of quantum statistics [8]. This limit is reached when the densities of quantum states are low, so that $\bar{n}_i << 1$ for all occupation numbers. In this case, expressions (1.147) and (1.148) give

$$e^{\beta(\mathcal{E}_i-\mu)} >> 1 \quad \text{so that} \quad \bar{n}_i \approx e^{-\beta(\mathcal{E}_i-\mu)} . \tag{1.170}$$

Consequently, we have obtained an approximate expression for the total number of particles,

$$N = \sum_i \bar{n}_i \approx e^{\beta\mu} \sum_i e^{-\beta\mathcal{E}_i} , \tag{1.171}$$

so that

$$e^{\beta\mu} \approx \frac{N}{\sum_j e^{-\beta\mathcal{E}_j}} \Longrightarrow \bar{n}_i = \frac{N\,e^{-\beta\mathcal{E}_i}}{\sum_j e^{-\beta\mathcal{E}_j}} . \tag{1.172}$$

Compare this expression with (1.153). The conclusion is that in the limit of low densities, *the occupational averages for quantum statistics are similar to those given by the Maxwell–Boltzmann distribution function.*

If the number of particles N is fixed, the grand partition function can be expressed as

$$\mathcal{Z} = \sum_{N'} Z(N', \beta)\, e^{\beta\mu N'}\, \delta_{N-N'} \approx Z(N)\, e^{\beta\mu N} . \tag{1.173}$$

Then, taking into account (1.146), we obtain for low densities

$$\ln \mathcal{Z} = \beta\mu N + \ln Z(N) \implies \ln Z(N) = -\beta\mu N + \ln \mathcal{Z}$$
$$= -\beta\mu N \pm \sum_i \ln(1 \pm e^{-\beta(\mathcal{E}_i - \mu)})$$

$$\approx -\beta\mu N \pm \sum_i (\pm) e^{-\beta(\mathcal{E}_i - \mu)} = -\beta\mu N + e^{\beta\mu} \sum_i e^{-\beta\mathcal{E}_i}. \qquad (1.174)$$

The last term is the classical expression for N so that $N \approx e^{\beta\mu} \sum_i e^{-\beta\mathcal{E}_i}$ if the density is low. Thus, (1.174) becomes

$$\ln Z(N) \approx -\beta\mu N + N = N - N \ln\left(\frac{N}{\sum_i e^{-\beta\mathcal{E}_i}}\right)$$

$$= N - N \ln N + N \ln\left(\sum_i e^{-\beta\mathcal{E}_i}\right) \approx \ln\left[\frac{1}{N!}\left(\sum_i e^{-\beta\mathcal{E}_i}\right)^N\right], \qquad (1.175)$$

where in the last step we have applied the Stirling formula. As a consequence, it is seen that the quantum statistics contains the factor $1/N!$ coming from the correct counting of states for indistinguishable particles.

1.5.2 The Grand Canonical Partition Function in the Continuum Limit

Now, we may understand why we have corrected, as done in (1.153), the formula for the partition function in (1.152). The classical limit of the partition function for quantum statistics is not exactly the partition function for the Maxwell–Boltzmann statistics. It includes the factor $1/N!$ that corrects the counting of microstates when the indistinguishability condition is considered.

Next, let us consider a three-dimensional particle enclosed in a box of volume V. Then, it seems natural to impose periodic boundary conditions for the values of each component of the momentum so that

$$k_i = \frac{2\pi}{L} n_i, \qquad (1.176)$$

where L is the length of the box edge and n_i is an integer, as required by the vanishing of the wave function at the extremes of the box. In the passage to the continuum, we may consider that the element of the density of states dn is then ($k = |\mathbf{k}|$)

$$dn = \frac{V}{(2\pi)^3} d\mathbf{k} = \frac{V}{2\pi^2} k^2 dk. \qquad (1.177)$$

Since the energy $\mathcal{E}_k$ corresponding to the value k is $\mathcal{E}_k = (\hbar k)^2/(2m)$, we have that the density of states in the continuum is given by

$$k = \left(\frac{2m\mathcal{E}}{\hbar^2}\right)^{1/2} \implies dk = \frac{1}{2}\left(\frac{2m}{\hbar^2}\right)^{1/2} \mathcal{E}^{-1/2}\,d\mathcal{E}. \tag{1.178}$$

Consequently,

$$dn = \left\{\frac{4\pi V}{(2\pi)^3}\left(\frac{2m}{\hbar^2}\right)^{3/2}\frac{1}{2}\right\}\mathcal{E}^{1/2}\,d\mathcal{E} = \rho(\mathcal{E})\,d\mathcal{E}, \tag{1.179}$$

where the expression for the energy density $\rho(\mathcal{E})$ is clear. We may include in (1.179) a degeneracy g due to the existence of internal degrees of freedom, like spin, etc. In this circumstance, (1.179) becomes

$$dn = \left\{\frac{4\pi V}{(2\pi)^3}\left(\frac{2m}{\hbar^2}\right)^{3/2}\frac{g}{2}\right\}\mathcal{E}^{1/2}\,d\mathcal{E}. \tag{1.180}$$

Let us use the symbol Ω to denote the free energy for a grand canonical ensemble. The free energy and the grand partition function are related as

$$\Omega = -\frac{1}{\beta}\ln\mathcal{Z} \iff \mathcal{Z} = e^{-\beta\Omega}. \tag{1.181}$$

In the thermodynamic definition of Ω, we have that $\Omega = -pV$. Therefore, the information on the equation of state of the system should come after the explicit expression for $\mathcal{Z}$ in the continuum. This can be done as follows: For fermions (sign $+$) and bosons (sign $-$), we have obtained the following expressions ($\alpha = -\beta\mu$):

$$\ln\mathcal{Z} = \pm\sum_i \ln(1 \pm e^{-(\alpha+\beta\mathcal{E}_i)}). \tag{1.182}$$

In the passage to the continuum, we have to integrate with respect to the density of states dn:

$$\ln\mathcal{Z} = \pm A_0 \int_0^\infty \ln(1 \pm e^{-(\alpha+\beta\mathcal{E}_i)})\,\mathcal{E}^{1/2}\,d\mathcal{E}, \tag{1.183}$$

where after (1.180), it is obvious that

$$A_0 = \frac{4\pi V}{(2\pi)^3}\left(\frac{2m}{\hbar^2}\right)^{3/2}\frac{g}{2}. \tag{1.184}$$

By using integration by parts on the integral in (1.183), we can obtain an interesting expression, namely,

$$\ln\mathcal{Z} = \pm A_0 \left\{\frac{2}{3}\mathcal{E}^{3/2}\ln(1 \pm e^{-(\alpha+\beta\mathcal{E})})\Big|_0^\infty \pm \frac{2\beta}{3}\int_0^\infty \frac{\mathcal{E}^{3/2}e^{-(\alpha+\beta\mathcal{E})}}{1 \pm e^{-(\alpha+\beta\mathcal{E})}}\,d\mathcal{E}\right\}. \tag{1.185}$$

The first term in the right-hand side of (1.185) vanishes, so that the final expression for the free energy in the grand canonical ensemble is

$$\Omega = -\frac{2}{3} A_0 \int_0^\infty \frac{\mathcal{E}^{3/2} e^{-(\alpha+\beta\mathcal{E})}}{1 \pm e^{-(\alpha+\beta\mathcal{E})}} d\mathcal{E}, \tag{1.186}$$

where the signs $+$ and $-$ stand for fermions and bosons, respectively. Let us make an estimation for the value of the integral in (1.186). First, let us divide numerator and denominator in the expression under the integral sign in (1.186) by $e^{-(\alpha+\beta\mathcal{E})}$. This yields

$$I_{3/2} = \int_0^\infty \frac{\mathcal{E}^{3/2}}{e^{(\alpha+\beta\mathcal{E})} \pm 1} d\mathcal{E}. \tag{1.187}$$

Then, let us use a new variable $x = \beta\mathcal{E}$. With this new variable, the integral (1.187) becomes

$$I_{3/2} = \left(\frac{1}{\beta}\right)^{5/2} \int_0^\infty \frac{x^{3/2}}{e^\alpha e^x \pm 1} dx = \left(\frac{1}{\beta}\right)^{5/2} \int_0^\infty \frac{x^{3/2} e^{-\alpha} e^{-x}}{1 \pm e^{-\alpha} e^{-x}} dx$$

$$= \frac{e^{-\alpha}}{\beta^{5/2}} \int_0^\infty x^{3/2} e^{-x} (1 \pm e^{-\alpha} e^{-x})^{-1} dx. \tag{1.188}$$

Since $\lambda = e^{-\alpha} \ll 1$, we have that

$$(1 \pm \lambda e^{-x})^{-1} = 1 \mp \lambda e^{-x} \pm \lambda^2 e^{-2x} \ldots . \tag{1.189}$$

If we insert (1.189) in (1.188) and keep the first three terms only, we have

$$I_{3/2} \approx \left(\frac{e^{-\alpha}}{\beta^{5/2}}\right) \int_0^\infty x^{3/2} e^{-x} (1 \mp e^{-\alpha} e^{-x} \pm e^{-2\alpha} e^{-2x}) dx$$

$$= \left(\frac{e^{-\alpha}}{\beta^{5/2}}\right) \left\{ \int_0^\infty x^{3/2} e^{-x} dx \mp e^{-\alpha} \int_0^\infty x^{3/2} e^{-2x} dx \pm e^{-2\alpha} \int_0^\infty x^{3/2} e^{-3x} dx \right\}. \tag{1.190}$$

Recall that

$$\Gamma(\nu) = \int_0^\infty e^{-x} x^{\nu-1} dx, \tag{1.191}$$

and use (1.191) in (1.190), we have that up to second order in λ

$$I_{3/2} \approx \left(\frac{e^{-\alpha}}{\beta^{5/2}}\right) \left[\Gamma\left(\frac{5}{2}\right) \mp e^{-\alpha} \frac{1}{2^{5/2}} \Gamma\left(\frac{5}{2}\right)\right] \tag{1.192}$$

and, in consequence

$$\Omega \approx -\frac{2}{3} A_0 \frac{1}{\beta^{5/2}} \Gamma\left(\frac{5}{2}\right) \left\{ e^{-\alpha} \left(1 \mp \frac{e^{-\alpha}}{2^{5/2}}\right) \right\}, \tag{1.193}$$

where the upper sign stands for fermions and the lower sign for bosons as before.

The total number of particles, at the same order of approximation, is given by

$$N = \int_0^\infty dn = A_0 \int_0^\infty \frac{\mathcal{E}^{1/2}}{e^{\alpha+\beta\mathcal{E}} \pm 1}\, d\mathcal{E} \approx A_0\, e^{-\alpha}\, \frac{1}{\beta^{3/2}}\, \Gamma\left(\frac{3}{2}\right)\left\{1 \mp \frac{e^{-\alpha}}{2^{3/2}}\right\}.$$
(1.194)

Equations (1.194) and (1.193) can be written, respectively, as

$$C_0 := \frac{N\,\beta^{3/2}}{A_0\,\Gamma(3/2)} = e^{-\alpha}\left(1 \mp \frac{e^{-\alpha}}{2^{3/2}}\right)$$
(1.195)

and

$$-\frac{\Omega\,\beta^{5/2}}{(2/3)A_0\,\Gamma(5/2)} = e^{-\alpha}\left(1 \mp \frac{e^{-\alpha}}{2^{5/2}}\right).$$
(1.196)

Taking into account that $\Gamma(3/2) = \sqrt{\pi}/2$, Eq. (1.195) yields

$$C_0 = \frac{2N\,\beta^{3/2}}{A_0\,\sqrt{\pi}}$$
(1.197)

For the Fermi–Dirac statistics, (1.197) reduces to

$$\lambda^2 - 2^{3/2}\,\lambda + 2^{3/2}\,C_0 = 0,$$
(1.198)

so that

$$\lambda = e^{-\alpha} = \sqrt{2}[1 \pm (1 - \sqrt{2}\,C_0)^{1/2}].$$
(1.199)

Because of (1.192), low values of $\lambda \ll 1$ correspond to small values of C_0. Therefore, the square root in the right-hand side of (1.199) can be expanded in Taylor series, which, up to second order, gives

$$(1 - \sqrt{2}\,C_0)^{1/2} = 1 - \frac{1}{2}(\sqrt{2}\,C_0) - \frac{1}{8}(\sqrt{2}\,C_0)^2 + \dots,$$
(1.200)

which implies that

$$e^{-\alpha} \approx \sqrt{2}\left\{1 \pm \left(1 - \frac{1}{\sqrt{2}}\,C_0 - \frac{1}{4}\,C_0^2\right)\right\} = C_0 + \frac{1}{2\sqrt{2}}\,C_0^2,$$
(1.201)

where only the solution with the lower sign is kept because $e^{-\alpha} > 0$.

It is straightforward to check that up to this order $e^{-\alpha}$, we have the following identity:

$$e^{-\alpha} - \frac{1}{2^{5/2}} e^{-2\alpha} = C_0 + C_0^2 \left[\frac{1}{2^{3/2}} - \frac{1}{2^{5/2}} \right] = C_0 + \frac{1}{4\sqrt{2}} C_0^2 . \tag{1.202}$$

Using (1.202) in (1.194), we have

$$-\frac{\Omega \beta^{5/2}}{(2/3)A_0 \Gamma(5/2)} = C_0 + \frac{1}{4\sqrt{2}} C_0^2 . \tag{1.203}$$

Consequently,

$$\Omega = -\frac{2}{3} A_0 \Gamma(5/2) \frac{1}{\beta^{5/2}} \left(\frac{N \beta^{3/2}}{A_0 \Gamma(3/2)} \right) \left(1 + \frac{N \beta^{3/2}}{A_0 \Gamma(3/2)} \frac{1}{4\sqrt{2}} \right)$$

$$= -\frac{2}{3} \frac{N}{\beta} \frac{\Gamma(5/2)}{\Gamma(3/2)} \left(1 + \frac{N \beta^{3/2}}{A_0 4\sqrt{2} \Gamma(3/2)} \right) = -Nk_BT \left(1 + \frac{N \beta^{3/2}}{A_0 4\sqrt{2} \Gamma(3/2)} \right), \tag{1.204}$$

where in the last identity, we have used that $\Gamma(5/2) = (3/2)\Gamma(3/2)$. Taking into account the expression for A_0 given in (1.184), we obtain an expression for the grand canonical free energy in the Fermi–Dirac (FD) statistics as

$$\Omega_{FD} \approx -Nk_BT \left(1 + \frac{N}{V} \frac{1}{(k_BT)^{3/2}} \frac{\hbar^3 \pi^{3/2}}{2g_s m^{3/2}} \right) . \tag{1.205}$$

Since $\Omega = -pV$, where p and V denote pressure and volume, respectively, relation (1.205) means that the pressure for the fermion gas is bigger than the pressure for a classical gas, for which $pV = Nk_BT$. This effect is due to the Pauli principle, according to which each energy level may only be occupied by a single particle. It can be interpreted as a repulsion, which increases the pressure respect to the classical gas.

Proceeding in an analogous manner, we obtain the corresponding expression for the *Bose–Einstein statistics* (BE)

$$\Omega_{BE} \approx -Nk_BT \left(1 - \frac{N}{V} \frac{1}{(k_BT)^{3/2}} \frac{\hbar^3 \pi^{3/2}}{2g_s m^{3/2}} \right) . \tag{1.206}$$

We see that the effect in the pressure for the Bose–Einstein gas is the inverse than in the Fermi–Dirac gas. In fact, the pressure for the BE gas is smaller than for its equivalent classical gas, an effect produced by the fact that bosons may occupy any energy level without restrictions [9].

1.5.2.1 Summary

- The evaluation of the partition function and average occupation numbers for quantum systems, with either integer (bosons) or half-integer spin (fermions), was given by introducing a variable number of particles (grand canonical ensemble) and a Lagrange multiplier (the chemical potential μ). In this situation, the average occupation number for the ith energy level is given by

$$\overline{n}_i = -\frac{1}{\beta}\left(\frac{\partial \ln \mathcal{Z}}{\partial \mathcal{E}_i}\right),$$

where $\mathcal{Z}$ is the grand partition function and $\mathcal{E}_i$ is the energy of the ith level.

- The expression given the fluctuations for average number of particles $\overline{n}_i$ is

$$\overline{(\Delta \overline{n}_i)^2} = -\frac{1}{\beta}\left(\frac{\partial \overline{n}_i}{\partial \mathcal{E}_i}\right).$$

- The correspondence between the grand canonical partition function $\mathcal{Z}$ and the grand canonical free energy (or grand potential) is

$$\mathcal{Z} = e^{-\beta\Omega} \iff \Omega = -\frac{1}{\beta}\ln \mathcal{Z}.$$

- We have used the so-called *maximal term approximation*, which is

$$\mathcal{Z}(\beta, \alpha) \approx Z(N)\, e^{-\alpha N}.$$

- Quantum distributions contain naturally the correction factor $1/N!$, which is associated to the indistinguishability of particles. In the Maxwell–Boltzmann statistics, this factor has to be introduced by hand, in order to avoid inconsistencies in the behavior of the entropy, such as the Gibbs paradox.

1.6   Some Features of Quantum Distributions

Let us recall the types of quantum distributions functions or quantum statistics. As we have seen, they can be classified as follows:

- *Distributions with a fixed number of particles*:
  - *Fermi–Dirac*. Corresponds to *fermions* or quantum particles with half-integer spin. Fermions obey the Pauli exclusion principle.
  - *Bose–Einstein*. Corresponds to *bosons* or quantum particles with integer spin. Bosons do not obey the exclusion principle, so that any quantum state may be occupied by an arbitrary number of bosons.

- Distribution functions without the constraint in the number of particles:
  - *Planck*. Valid for massless particles.

The classical limit of the quantum distributions gives the *corrected* Maxwell–Boltzmann distribution. In the sequel, we study the statistical behavior of fermions and bosons and recover the classical limit of their distributions.

1.6.1 Fermions

To begin with, the average occupation number has been given in (1.149). This equation is valid for non-zero $(T \neq 0)$ temperature. Then, in terms of the parameter $\alpha = -\beta\mu$, the expression (1.149) can be written as

$$\bar{n}_i = \frac{1}{e^{\alpha + \beta \mathcal{E}_i} + 1}. \tag{1.207}$$

Note that, in the limit of zero temperature, we have

$$\lim_{T \mapsto 0^+} \bar{n}_i = \theta(\mathcal{E}_i - \mu), \tag{1.208}$$

where $\theta(x)$ is the step function:

$$\theta(x) := \begin{cases} 1 \text{ if } x < 0 \\ 0 \text{ if } x > 0 \end{cases}. \tag{1.209}$$

Then, formula (1.208) shows that in the limit $T = 0$ all the energy levels with energies $\mathcal{E}_i < \mu$ are occupied, while all others are empty. Therefore, the value of μ provides the beginning of the energy gap between occupied and empty energy levels. The parameter μ is known as the *Fermi level*, which can be determined as a solution of the equation

$$N = \sum_i \bar{n}_i(T = 0) = \sum_{i : \mathcal{E}_i < \mu} 1, \tag{1.210}$$

N being the total number of particles. The index i denotes the energy levels with the corresponding degeneracy.

In the continuum limit, the total number of particles is obtained by replacing the sum in (1.210) by an integral with the density $\rho(\mathcal{E}) \, d\mathcal{E} = A_0 \mathcal{E}^{1/2} \, d\mathcal{E}$ with A_0 as in (1.192):

$$N = A_0 \int_0^\infty \frac{\mathcal{E}^{1/2}}{e^{\alpha + \beta \mathcal{E}} + 1} \, d\mathcal{E}. \tag{1.211}$$

Taking the limit of zero temperature, the function $(e^{\alpha + \beta \mathcal{E}} + 1)^{-1}$ becomes the step function, $\theta(\mathcal{E}_i - \mu)$, so that (1.115) reads

$$N = A_0 \int_0^{\mu(T=0)} \mathcal{E}^{1/2} \, d\mathcal{E} = \frac{2}{3} A_0 \mu^{3/2} (T = 0). \tag{1.212}$$

Taking into account the explicit expression for A_0 given in (1.184), we can determine the value of the Fermi level from (1.125) as

$$\mu(T = 0) = \left(\frac{3}{2}\frac{N}{A_0}\right)^{2/3} = \frac{\hbar^2}{2m}\left(6\pi^2\frac{N}{g_s V}\right)^{2/3} = \frac{\hbar^2}{2m}k_F^2. \quad (1.213)$$

Formula (1.213) defines the *value of the Fermi momentum, k_F at $T = 0$*.

In general, we have to replace $\mathcal{E}^{1/2}$, which is valid for a free particle (or plane waves), by a certain function $f(\mathcal{E})$ in (1.212), so that the total number of particles is

$$N = A_0 \int_0^\infty \frac{f(\mathcal{E})}{1 + e^{\beta(\mathcal{E}-\mu)}}\,d\mathcal{E} = A_0\,I(\mu), \quad (1.214)$$

where

$$I(\mu) = \int_0^\infty \frac{f(\mathcal{E})}{1 + e^{\beta(\mathcal{E}-\mu)}}\,d\mathcal{E} = \int_0^\mu \frac{f(\mathcal{E})}{1 + e^{\beta(\mathcal{E}-\mu)}}\,d\mathcal{E} + \int_\mu^\infty \frac{f(\mathcal{E})}{1 + e^{\beta(\mathcal{E}-\mu)}}\,d\mathcal{E}$$
$$= \int_0^\mu f(\mathcal{E})\,d\mathcal{E} - \int_0^\mu \frac{f(\mathcal{E})}{1 + e^{-\beta(\mathcal{E}-\mu)}}\,d\mathcal{E} + \int_\mu^\infty \frac{f(\mathcal{E})}{1 + e^{\beta(\mathcal{E}-\mu)}}\,d\mathcal{E}. \quad (1.215)$$

Let $F(\mathcal{E})$ be a primitive of $f(\mathcal{E})$, i.e., $F'(\mathcal{E}) = f(\mathcal{E})$, where the prime denotes derivation with respect to $\mathcal{E}$. Then, by performing the change of variables $x := \beta(\mathcal{E} - \mu)$ in the third integral in (1.215) and $x := -\beta(\mathcal{E} - \mu)$ in the second integral in (1.215), we have

$$I(\mu) = F(\mathcal{E})\Big|_0^\mu - \frac{1}{\beta}\int_0^{\beta\mu} \frac{f(\mu - x/\beta)}{1 + e^x}\,dx + \frac{1}{\beta}\int_0^\infty \frac{f(\mu + x/\beta)}{1 + e^x}\,dx. \quad (1.216)$$

For low temperature, $\beta\mu \gg 1$, the following approximations are valid:

$$f(\mu + x/\beta) \approx f(\mu) + f'(\mu)(x/\beta), \quad (1.217)$$
$$f(\mu - x/\beta) \approx f(\mu) - f'(\mu)(x/\beta). \quad (1.218)$$

Thus, when $\beta\mu \gg 1$, the upper limit in the first integral in (1.216) can be replaced by infinity. Then, using (1.217) and (1.218), one gets

$$I(\mu) = F(\mathcal{E})\Big|_0^\mu + \frac{2f'(\mu)}{\beta^2}\int_0^\infty \frac{x}{1 + e^x}\,dx = F(\mathcal{E})\Big|_0^\mu + \frac{\pi^2}{6}(k_B T)^2 f'(\mu), \quad (1.219)$$

since the value of the integral in (1.219) is $\pi^2/12$.

Then, for free particles we have that $f(\mathcal{E}) = \mathcal{E}^{1/2}$, $f'(\mathcal{E}) = \frac{1}{2}\mathcal{E}^{-1/2}$ and $F(\mathcal{E}) = \frac{2}{3}\mathcal{E}^{3/2}$. This gives along with (1.219) and (1.214),

$$N \approx A_0\left\{\frac{2}{3}\mu^{3/2} + \frac{\pi^2}{6}(k_B T)^2\frac{1}{2\mu^{1/2}}\right\}, \quad (1.220)$$

which is the relation between the total number of particles and μ at $T \neq 0$ (with $\beta\mu >> 1$).

For low temperatures, a first approximation for μ is $\mu(T = 0)$. Small corrections coming from the above expression for N add as

$$\frac{2}{3}\mu^{3/2}(T) \approx \frac{2}{3}\mu^{3/2}(T = 0) - \frac{\pi^2}{6}(k_B T)^2 \frac{1}{2\mu^{1/2}(T = 0)}, \tag{1.221}$$

from where

$$\mu^{3/2}(T) \approx \mu^{3/2}(T = 0)\left(1 - \frac{\pi^2}{12}\left(\frac{k_B T}{\mu(T = 0)}\right)^2\right), \tag{1.222}$$

which gives an approximate relation for the Fermi energy $\mu(T)$ with T at low temperatures.

A similar calculation can be performed for the total energy:

$$E = A_0 \int_0^\infty \frac{\mathcal{E}\,f(\mathcal{E})}{1 + e^{\beta(\mathcal{E}-\mu)}}\,d\mathcal{E} = A_0 \int_0^\infty \frac{\mathcal{E}^{3/2}}{1 + e^{\beta(\mathcal{E}-\mu)}}\,d\mathcal{E}. \tag{1.223}$$

At low temperatures, this gives

$$E \approx A_0 \left\{\frac{2}{5}\mu^{5/2} + \frac{\pi^2}{6}(k_B T)^2 \frac{3}{2}\mu^{1/2}\right\}. \tag{1.224}$$

We may write this formula in terms of T and $\mu(T = 0)$:

$$E \approx A_0 \left[\frac{2}{5}\left(\mu(T = 0)\left(1 - \frac{\pi^2}{12}\left(\frac{k_B T}{\mu(T = 0)}\right)^2\right)\right)^{5/2}\right.$$

$$\left. + \frac{\pi^2}{6}(k_B T)^2 \frac{3}{2}\mu^{1/2}(T = 0)\left(1 - \frac{\pi^2}{12}\left(\frac{k_B T}{\mu(T = 0)}\right)^2\right)^{1/2}\right]$$

$$\approx A_0 \left[\frac{2}{5}\mu^{5/2}(T = 0)\left(1 - \frac{\pi^2}{12}\frac{5}{2}\left(\frac{k_B T}{\mu(T = 0)}\right)^2\right) + \frac{3}{2}\mu(T = 0)\frac{\pi^2}{6}(k_B T)^2\right]$$

$$= \left[A_0 \frac{2}{5}\mu^{5/2}(T = 0) + A_0 \frac{\pi^2}{6}(k_B T)^2 \mu^{1/2}(T = 0)\right] = E_0 + aT^2. \tag{1.225}$$

From the second line in (1.225), we obtain the third by using the Taylor series for $(1 - x)^r$ and keeping the two first terms only. Then using the first identity in (1.213), relation (1.225) becomes

$$E = \frac{2}{5}\frac{1}{A_0^{2/3}}\left(\frac{3}{2}N\right)^{5/3} + A_0 \frac{\pi^2}{6}(k_B T)^2 \left(\frac{3}{2}\frac{N}{A_0}\right)^{1/3}$$

$$= E_0 + a(k_B T2)^2. \tag{1.226}$$

This readily gives

$$E_0 = \frac{3}{5} N \mu(T = 0)$$

$$a = \frac{\pi^2}{4} \frac{N}{\mu(T = 0)} . \tag{1.227}$$

This formula is useful to obtain the average energy per particle $\overline{E}$ at low temperature as

$$\overline{E} := \frac{E}{N} \approx \frac{3}{5} \mu(T = 0) + \frac{\pi^2}{4} \frac{1}{\mu(T = 0)} (k_B T)^2 . \tag{1.228}$$

Finally, the specific heat at constant volume per particle can be easily obtained as

$$\frac{C_V}{N} = \frac{d\overline{E}}{dT} = \frac{k_B}{\mu(T = 0)} \frac{\pi^2}{2} k_B T . \tag{1.229}$$

We shall return to the case of Fermi–Dirac distributions later on.

1.6.2 Bosons

Let us recapitulate on the concepts advanced for the Bose–Einstein quantum statistics. The average occupation number is ($\alpha = -\beta \mu$)

$$\overline{n}_i = \frac{1}{e^{\alpha + \beta \mathcal{E}_i} - 1} . \tag{1.230}$$

Next, consider low temperatures and the passage to the continuum. The total number of particles is

$$N = A_0 \int_0^\infty \frac{\mathcal{E}^{1/2}}{e^{\alpha + \beta \mathcal{E}} - 1} d\mathcal{E} \tag{1.231}$$

and the total energy is

$$E = A_0 \int_0^\infty \frac{\mathcal{E}^{3/2}}{e^{\alpha + \beta \mathcal{E}} - 1} d\mathcal{E} . \tag{1.232}$$

In order to give explicit expressions for N and $\mathcal{E}$, we need to calculate integrals of the following kind:

$$I_r(\alpha) = \int_0^\infty \frac{\mathcal{E}^r}{e^{\alpha + \beta \mathcal{E}} - 1} d\mathcal{E} . \tag{1.233}$$

Using a new variable x defined as $x = \sqrt{\beta \mathcal{E}}$, (1.233) becomes

$$I_r(\alpha) = \frac{2}{\beta^{r+1}} \int_0^\infty \frac{x^{2r+1}}{e^\alpha e^{x^2} - 1} dx = \frac{2}{\beta^{r+1}} \int_0^\infty \frac{x^{2r+1}}{e^\alpha e^{x^2}(1 - e^{-\alpha} e^{-x^2})} dx$$

$$= \frac{2}{\beta^{r+1}} \int_0^\infty x^{2r+1} e^{-\alpha} e^{-x^2} (1 - e^{-\alpha} e^{-x^2})^{-1} dx . \tag{1.234}$$

Next, we use Taylor series in (1.234) for $(1 - e^{-\alpha} e^{-x^2})^{-1}$ to get the following expression ($\lambda = e^{-\alpha}$):

$$I_r(\alpha) = \frac{2}{\beta^{r+1}} \int_0^\infty \lambda e^{-x^2} \left(\sum_{n=0}^\infty \lambda^n e^{-nx^2} \right) x^{2r+1} \, dx$$

$$= \frac{2}{\beta^{r+1}} \sum_{n=1}^\infty \lambda^n \left\{ \int_0^\infty x^{2r+1} e^{-nx^2} \, dx \right\}. \tag{1.235}$$

In order to solve the integral in (1.235), we perform the change of variables $t = nx^2$ and define $\nu := 2r + 1$. Then, this integral gives

$$\int_0^\infty x^\nu e^{-nx^2} \, dx = \int_0^\infty \frac{1}{2n} \left(\frac{t}{n} \right)^{-1/2} \left(\frac{t}{n} \right)^{(2r+1)/2} e^{-t} \, dt$$

$$= \frac{1}{2n^{r+1}} \int_0^\infty t^r e^{-t} \, dt. \tag{1.236}$$

Note that the integral in the second row of (1.236) can be expressed as the value of the Gamma function at $r + 1$, $\Gamma(r + 1)$. Therefore, (1.235) takes the form

$$I_r(\lambda) = \frac{\Gamma(r + 1)}{\beta^{r+1}} \sum_{n=1}^\infty \frac{\lambda^n}{n^{r+1}}. \tag{1.237}$$

The sum of the series,

$$\xi_{r+1}(\lambda) := \sum_{n=1}^\infty \frac{\lambda^n}{n^{r+1}}, \tag{1.238}$$

defines a function which is known as the *polylogarithm*. Here, λ is a fixed number. When we choose $\lambda = 1$, the polylogarithm becomes the Riemann zeta function, so that the polylogarithm can be looked as a generalization of this function. A compact expression for (1.237) is

$$I_r(\lambda) = \frac{\Gamma(r + 1)}{\beta^{r+1}} \xi_{r+1}(\lambda). \tag{1.239}$$

The values of the total number N of particles and the total energy E can be obtained by choosing $r = 1/2$ and $r = 3/2$, respectively. See Eqs. (1.231) and (1.232). We have the following expressions:

$$N = A_0 \frac{\Gamma(3/2)}{\beta^{3/2}} \xi_{3/2}(\lambda) \tag{1.240}$$

and

$$E = A_0 \frac{\Gamma(5/2)}{\beta^{5/2}} \xi_{5/2}(\lambda). \tag{1.241}$$

Then, using $\Gamma(1/2) = \sqrt{\pi}$, the properties of the Gamma function and the value of A_0 given in (1.192), we obtain from (1.240)

$$N = V \frac{g_s \, m^{3/2}}{\sqrt{2}\pi^2 \hbar^3} \frac{\pi^{1/2}}{2\beta^{3/2}} \xi_{3/2}(\lambda) = V \, g_s \left(\frac{m}{2\pi\hbar^2}\right)^{3/2} (k_B T)^{3/2} \xi_{3/2}(\lambda) . \quad (1.242)$$

The density, $\rho = N/V$, for a system of N bosons with mass m in a finite volume V, in this limit reads

$$\rho = g_s \left(\frac{m}{2\pi\hbar^2}\right)^{3/2} (k_B T)^{3/2} \xi_{3/2}(\lambda) . \quad (1.243)$$

The expression for the total energy E is after (1.241)

$$E = V \, g_s \left(\frac{m}{2\pi\hbar^2}\right)^{3/2} \frac{3}{2} (k_B T)^{5/2} \xi_{5/2}(\lambda) . \quad (1.244)$$

Replacing in (1.244) V by $V = N\rho$ and applying (1.243), we have

$$E = \frac{3}{2} N k_B T \, \frac{\xi_{5/2}(\lambda)}{\xi_{3/2}(\lambda)} . \quad (1.245)$$

Although the previous formulae have been obtained for low temperatures, one may be interested in the behavior of the quotient between the polylogarithms in (1.245) for very high temperatures. In this case, $\lambda \approx 1$, and because of the properties of the polylogarithm give, the ratio between $\xi_{5/2}(\lambda)$ and $\xi_{3/2}(\lambda)$ is

$$\frac{\xi_{5/2}(\lambda)}{\xi_{3/2}(\lambda)} \approx 1 . \quad (1.246)$$

One might use this last relation in order to argue that in the limit of high temperatures, we recover the classical limit

$$E \approx \frac{3}{2} N k_B T . \quad (1.247)$$

1.7 Bose–Einstein Condensation

As a matter of fact for systems of bosons, the relation (1.240) should be valid to account for excited states. The total number of particles, fixed at the value N, is a sum of particles occupying the ground state, N_{gs}, plus particles in excited states, N_{exc}:

$$N = N_{gs} + N_{exc} . \quad (1.248)$$

Let us define T_0 as the *minimal temperature at which all particles occupy excited states*, so that the ground state is empty. We call T_0 the *transition temperature*, and for it

$$N = N_{exc} = A_0 \int_0^\infty \frac{\mathcal{E}^{1/2}}{e^{(\mathcal{E}-\mu)/(k_B T_0)} - 1} \, d\mathcal{E} = A_0 \frac{\sqrt{\pi}}{2} k_B^{3/2} \xi_{3/2}(1) \, T_0^{3/2} . \quad (1.249)$$

This identity provides a precise formula to evaluate T_0:

$$k_B T_0 = \left\{ \frac{2N}{A_0 \sqrt{\pi}} \frac{1}{\xi_{3/2}(\lambda = 1)} \right\}^{2/3} . \quad (1.250)$$

For temperatures below T_0 ($T < T_0$), the number of particles occupying excited states should be much smaller than the number of particles occupying the ground state. In fact, the integral for the number of particles becomes maximal for $\mu = 0$. This is

$$N = A_0 \int_0^\infty \frac{\mathcal{E}^{1/2}}{e^{(\mathcal{E})/(k_B T)} - 1} \, d\mathcal{E} = A_0 \frac{\sqrt{\pi}}{2} k_B^{3/2} \xi_{3/2}(\lambda = 1) \, T^{3/2} . \quad (1.251)$$

Therefore for $T < T_0$, we have the following relation:

$$\frac{N_{exc}}{N} = \left(\frac{T}{T_0} \right)^{3/2} \iff N_{exc} = N \left(\frac{T}{T_0} \right)^{3/2} . \quad (1.252)$$

Consequently, the number of particles in the ground state is

$$N_{gs} = N - N_{exc} = N \left[1 - \left(\frac{T}{T_0} \right)^{3/2} \right] . \quad (1.253)$$

For $T = 0$ all particles are in the ground state and for low temperatures $N_{gs} \approx N$. This is an effect called the *Bose–Einstein condensation*, and the resulting N-particle state is named a *Bose–Einstein condensate*. It is important to remark that a Bose–Einstein condensation *is not a spatial condensation*, but a condensation in the space of momenta. In a reference frame in which the ground state has zero energy, it means that all particles in the condensate have (approximately) zero momenta. The effect is due purely to the statistics. It depends on the dimensions of the system and should not be taken as a phase transition, since no interaction between the particles is present in the Hamiltonian, which is the case for free particles.

1.7.1 Massive Fermions in the Relativistic Limit

In the case of massive fermions, like electrons, the energy of a single state is given by the well-known relation between rest mass, m, and momentum, p, namely, $\epsilon = \sqrt{(pc)^2 + m^2c^4}$, where c is the speed of light. The differential element in phase space will then be

$$d\Omega(p, q) = \frac{d^3p\, d^3q}{(2\pi\hbar)^3} g_s$$

$$= \frac{V}{\pi^2\hbar^3} p^2\, dp \tag{1.254}$$

for the spin degeneracy $g_s = 2$. By introducing the variable $x = \epsilon - mc^2$, integrals of the type

$$I(k) = \int dp\, p^2 n(\epsilon) \left(\frac{\epsilon}{mc^2}\right)^k \tag{1.255}$$

where $n(\epsilon)$ is the average occupation number for fermions, that is, $n(\epsilon) = 1/(1 + exp(\epsilon - \mu)/kT)$, being μ the chemical potential and T the temperature, transform like

$$I(k) = \frac{1}{2}(2mkT)^{3/2} \int_0^\infty dw\, w^{1/2} (1 + \frac{1}{2}\gamma w)^{1/2} (1 + \gamma w)^{k+1}/(1 + e^{(w-\eta)}), \tag{1.256}$$

where we have made use of the following replacements:

$$w = x/kT \qquad \gamma = kT/mc^2 \qquad \eta = (\mu - mc^2)/kT. \tag{1.257}$$

The integrals (1.256) may be further reduced to integrals of the type

$$F(r) = \int_0^\infty dw\, w^r (1 + \frac{1}{2}\gamma w)^{1/2}/(1 + e^{(w-\eta)}). \tag{1.258}$$

Some useful integrals are the following:

$$I(k = 0) = (1/2)(2mkT)^{3/2}(F(1/2) + \gamma F(3/2))$$
$$I(k = -1) = (1/2)(2mkT)^{3/2} F(1/2)$$
$$I(k = 1) = (1/2)(2mkT)^{3/2}(F(1/2) + 2\gamma F(3/2) + \gamma^2 F(5/2)). \tag{1.259}$$

In terms of these integrals, the particle density ($\rho = N/V$) and the energy density ($U = E/V$) of relativistic free fermions of rest mass m are written as

$$\rho = \frac{1}{\pi^2\hbar^3} I(k = 0)$$

$$U = \frac{1}{\pi^2\hbar^3} I(k = 1), \tag{1.260}$$

respectively.

1.7.2 Massless Relativistic Fermions

The expressions given above are further reduced to much simpler forms for the case of ultrarelativistic $pc >> mc^2$ or massless fermions, for which $\epsilon = pc$. Proceeding in the same manner as before, the corresponding expressions for the particle density and energy density are written as

$$\rho = \frac{(k_B T)^3}{\pi^2 \hbar^3 c^3}(7/8)\Gamma(3)$$

$$U = \frac{(k_B T)^4}{\pi^2 \hbar^3 c^3}(15/16)\Gamma(4) \tag{1.261}$$

with $\Gamma(n) = \int_0^\infty dw\, w^{(n-1)} e^{(-w)} = (n-1)!$. From these expressions we can write the relationship between energy density and particle density, for massless (or ultra-relativistic) fermions at finite temperature T, that is,

$$U = (90/28)\, \rho\, kT \approx 3.214\, \rho\, kT. \tag{1.262}$$

The results given in the previous two subsections are of some use in astrophysical applications.

1.8 Summary

- In terms of the temperature, the number of (excited) bosons and the total value of the energy are given by Eqs. (1.241) and (1.241), respectively.
- There is a minimal temperature T_0 at which all states are excited. The explicit value of T_0 is given in (1.250). For very low temperatures below T_0, most of the particles are in the ground state, a situation known as Bose–Einstein condensation.
- When the temperature increases up to T_0, the ground state loses particles. At T_0 the ground state becomes empty.
- When the temperature T is higher than T_0, $T > T_0$, the ground state remains empty. At very high temperatures, the Bose–Einstein distribution becomes the classical Maxwell–Boltzmann.
- The Bose–Einstein condensation **is not** a phase transition. Phase transitions are characterized by singularities of the free energy in terms of parameters like the temperature, density, etc. The Bose–Einstein condensation is just an inversion of the occupation number (or the population of the ground state).
- The specific heat at constant volume has a dependence with temperature of the order of $T^{3/2}$ up to $T = T_0$ and then tends smoothly to a constant for higher temperatures. In any case, the value of C_V is finite, although it reaches a maximum, at T_0. This again gives an evidence of the absence of phase transition at $T = T_0$.

The Role of Dynamics in Statistical Mechanics

2

In this chapter, we shall address the question concerning the relationships between the equations of motion in Lagrangian mechanics with the probabilistic interpretation of statistical mechanics. The motivation is obvious, since the probabilistic conception of statistical mechanics is based on the knowledge of the exact energy levels of the system under study, something which in practice is restricted to very few examples. In other words, in dealing with real physical systems, one is forced to make approximations in order to construct the energy spectrum and to determine their degeneracy. In both classical and quantum mechanics, the way to represent physical systems often goes by using perturbation theory. Not so often one finds in the literature a perturbative approach applied to statistical mechanics. This may be due to the impossibility of defining uniquely an expansion parameter. In consequence, following the Feynman approach, it may be worthy to rephrase the statistical approach presented in the previous chapter, in terms of amplitudes rather than probabilities. Then, all the machinery currently used in field theory, such as equations of motions and expectation values, can be applied to find out statistical observables . The same notions have been applied to mechanical and statistical systems by Bogolubov [10].

2.1 Statistic Distributions

In the sequel, we are going to discuss some features concerning statistical distributions from a geometrical point of view. We shall start with the classical aspects to go on with the quantum case later.

2.1.1 The Liouville Theorem

Let us consider a classical system of particles with s degrees of freedom, so that we have s positions and s conjugate momenta. Assume that the distribution function for this ensemble is $\rho(p, q)$, where $p = (p_1, \ldots, p_s)$ and $q = (q_1, \ldots, q_s)$. The system motion is Hamiltonian, i.e., it is governed by a Hamiltonian $H(p, q)$, which does not depend explicitly on time. Assume that the points evolve in time, or equivalently, are in a motion governed by $H(p, q)$. The total derivative of the density $\rho(p, q)$ with respect to time is

$$\frac{d\rho}{dt} = \frac{\partial \rho}{\partial t} + \sum_{i=1}^{s} \frac{\partial \rho}{\partial q_i} \dot{q}_i + \frac{\partial \rho}{\partial p_i} \dot{p}_i, \tag{2.1}$$

where the dot above a variable denotes derivative with respect to time. For each coordinate $i = 1, 2, \ldots, s$, we make use of the Hamilton equations of motion, which are

$$\dot{q}_i = \frac{\partial H}{\partial p_i}, \qquad \dot{p}_i = -\frac{\partial H}{\partial q_i}, \qquad i = 1, 2, \ldots, s, \tag{2.2}$$

so that (2.1) becomes

$$\frac{d\rho}{dt} = \frac{\partial \rho}{\partial t} + \sum_{i=1}^{s} \left\{ \frac{\partial \rho}{\partial q_i} \frac{\partial H}{\partial p_i} - \frac{\partial \rho}{\partial p_i} \frac{\partial H}{\partial q_i} \right\} = \frac{\partial \rho}{\partial t} + \{\rho, H\}. \tag{2.3}$$

If A and B are functions of (p, q), the Poisson bracket $\{A, B\}$ of A and B is defined as

$$\{A, B\} := \sum_{i=1}^{s} \left\{ \frac{\partial A}{\partial q_i} \frac{\partial B}{\partial p_i} - \frac{\partial B}{\partial p_i} \frac{\partial A}{\partial q_i} \right\}. \tag{2.4}$$

Note that $\{A, B\} = -\{B, A\}$, Equation (2.3) is known as the *Liouville equation*. It is related with the continuity equation that obeys all densities due to the law of conservation of some quantity, like mass or energy. Assume that an ensemble of particles is moving on phase space under the assumption of no loss of mass (in our case, no loss of particles). Let $\mathbf{u} := (\dot{q}_i, \dot{q}_2, \ldots, \dot{q}_s, \dot{p}_1, \dot{p}_2, \ldots, \dot{p}_s)$ be the velocity of a point in phase space due to the motion in the system of particles produced by H. Then, the *flux* $\mathbf{j}$ is defined as $\mathbf{j} := \rho \mathbf{u}$, where ρ is the density $\rho(p, q)$. Since the total mass is a conserved quantity, ρ and $\mathbf{j}$ have to satisfy the *continuity equation*

$$\frac{\partial \rho}{\partial t} + \nabla \cdot \mathbf{j} = 0. \tag{2.5}$$

Since

$$\nabla = \left(\frac{\partial}{\partial q_1}, \ldots, \frac{\partial}{\partial q_s}, \frac{\partial}{\partial p_1}, \ldots, \frac{\partial}{\partial p_s} \right), \tag{2.6}$$

we have that

$$
\nabla \cdot \mathbf{j} = \sum_{i=1}^{s} \left\{ \frac{\partial(\rho \, \dot{q}_1)}{\partial q_1} + \cdots \frac{\partial(\rho \, \dot{q}_s)}{\partial q_s} + \frac{\partial(\rho \, \dot{p}_1)}{\partial p_1} + \cdots + \frac{\partial(\rho \, \dot{p}_s)}{\partial p_s} \right\}
$$

$$
= \sum_{i=1}^{s} \left\{ \frac{\partial \rho}{\partial q_1} \, \dot{q}_1 + \cdots \frac{\partial \rho}{\partial q_s} \, \dot{q}_s + \frac{\partial \rho}{\partial p_1} \, \dot{p}_1 + \cdots + \frac{\partial \rho}{\partial p_s} \, \dot{p}_s \right\}
$$

$$
+ \rho \sum_{i=1}^{s} \left\{ \frac{\partial \dot{q}_1}{\partial q_1} + \cdots \frac{\partial \dot{q}_s}{\partial q_s} + \frac{\partial \dot{p}_1}{\partial p_1} + \cdots \frac{\partial \dot{p}_s}{\partial p_s} \right\}. \tag{2.7}
$$

Look at the last row in (2.7). For each $i = 1, 2, \ldots, s$, we have the following terms:

$$
\frac{\partial \dot{q}_i}{\partial q_i} = \frac{\partial}{\partial q_i} \, \dot{q}_i = \frac{\partial}{\partial q_i} \frac{\partial H}{\partial p_i} = \frac{\partial^2 H}{\partial q_i \partial p_i} \tag{2.8}
$$

and

$$
\frac{\partial \dot{p}_i}{\partial p_i} = \frac{\partial}{\partial p_i} \, \dot{p}_i = -\frac{\partial}{\partial p_i} \frac{\partial H}{\partial q_i} = -\frac{\partial^2 H}{\partial p_i \partial q_i}, \tag{2.9}
$$

where we have applied the Hamilton equations (2.2). Since for all $i = 1, 2, \ldots, s$,

$$
\frac{\partial^2 H}{\partial q_i \partial p_i} = \frac{\partial^2 H}{\partial p_i \partial q_i}, \tag{2.10}
$$

we conclude that the last row in (2.7) vanishes. Then, using the Hamilton equations in the second row in (2.7), we finally have that

$$
\nabla \cdot \mathbf{j} = \{\rho, H\}, \tag{2.11}
$$

so that the continuity equation (2.5) takes the form

$$
\frac{\partial \rho}{\partial t} + \{\rho, H\} = 0. \tag{2.12}
$$

Going back to (2.3), we conclude that the density is constant with time along any trajectory:

$$
\boxed{\frac{d\rho}{dt} = 0}. \tag{2.13}
$$

From (2.12) and (2.13), we write

$$
\frac{\partial \rho}{\partial t} + \nabla \cdot \mathbf{j} = 0, \tag{2.14}
$$

which is the continuity equation common in fluid mechanics and electromagnetism, among other cases where the fluid is assumed to be non-compressible.

2.1.2   The Role of First Integrals

The Hamilton equations represent a system of $2s$ first-order differential equations of the form $\dot{x}_i = F(x_1, \ldots, x_{2s})$, $i = 1, 2, \ldots, 2s$, which is in general non-linear. In order to reduce the number of integrals and obtain constants of motion, we need first integrals. A first integral is a function $G(x_1, \ldots, x_{2s})$ that is constant along any trajectory, i.e., $G(x_1(t), \ldots, x_{2s}(t)) = G_0$, for all values of t [11].

For Hamiltonian systems and when the Hamiltonian does not depend explicitly on time, there are seven first integrals, which are the total energy, $E(p, q)$, the three components of the linear momentum, $\mathbf{P}(p, q)$, and the three components of the angular momentum, $\mathbf{J}(p, q)$. Thus, along any trajectory, we have

$$E(p(t), q(t)) = E_0 , \qquad \mathbf{P}(p(t).q(t)) = \mathbf{P}_0 , \qquad \mathbf{J}(p(t), q(t)) = \mathbf{J}_0 . \qquad (2.15)$$

Enforcing these constraints, the density of states in phase space takes the form

$$\rho(p, q) = \text{constant } \delta(E - E_0) \, \delta(\mathbf{P} - \mathbf{P}_0) \, \delta(\mathbf{J} - \mathbf{J}_0) . \qquad (2.16)$$

2.1.3   The Quantum Case

Let ψ be a pure state represented as a vector in a Hilbert space $\mathcal{H}$. If $\{\psi_n\}$ is an orthonormal basis in $\mathcal{H}$, then $\psi = \sum_n c_n \psi_n$, where $c_n = \langle \psi_n | \psi \rangle$, where $\langle - | - \rangle$ denotes the Hilbert space scalar product. Let us assume that the basis $\{\psi_n\}$ is formed by eigenvectors of the Hamiltonian (such as happens for the harmonic oscillator), i.e., $H\psi_n = E_n \psi_n$ for all values $n = 0, 1, 2, \ldots$. As well known, the time evolution of ψ is given by

$$U(t)\psi = e^{-iHt/\hbar}\psi = \sum c_n e^{-iHt/\hbar}\psi_n = \sum_n c_n e^{-itE_n/\hbar}\psi_n = \sum_n c_n(t)\psi_n ,$$
$$(2.17)$$

so that

$$c_n(t) = e^{-itE_n/\hbar} c_n , \qquad n = 0, 1, 2, \ldots . \qquad (2.18)$$

Then, if we call $\omega_{nm}(t) := c_m^*(t)c_n(t)$, we have that

$$\frac{d}{dt}\omega_{nm}(t) = \frac{i}{\hbar}(E_m - E_n)\,\omega_{nm}(t) . \qquad (2.19)$$

It is often convenient to use the notation $|n\rangle$ for ψ_n and also omit the dependence on t for $\omega_{nm}(t)$. This will simplify our expressions and will be used whenever it would not be cause for confusion. We also write $H_{ln} := \langle l|H|n \rangle = E_n \langle l|n \rangle = E_n \delta_{ln}$, where δ_{ln} is the Kronecker delta. Using this notation, it is straightforward that

$$E_n \omega_{mn} = \sum_l \omega_{ml} \langle l|H|n \rangle \implies (E_n - E_m)\omega_{mn} = \sum_l (\omega_{ml} H_{ln} - H_{ml} \omega_{ln}).$$

(2.20)

In the basis $\{|n\rangle\} \equiv \{\psi_n\}$, H_{nm} would play the role of *matrix entries* for the infinite matrix representing the Hamiltonian H. If we denote by ω, the operator uniquely defined by $\omega_{mn} = \langle m|\omega|n \rangle$, equation (2.20) can be written as

$$\frac{d}{dt}\omega = -\frac{i}{\hbar}[H, \omega] \iff i\hbar \frac{d}{dt}\omega = [H, \omega].$$

(2.21)

As a matter of fact, (2.21) is a particular case of the Liouville equation for quantum states. If ρ represents the operator for a quantum state, in general, a mixture, and H is a time-independent Hamiltonian given the time evolution of a quantum system, the variation with time of the observable ρ is given by

$$i\hbar\dot{\rho}(t) = [H, \rho],$$

(2.22)

where the dot above a variable means the total derivative of this variable with respect to time. As usual, $[A, B] = AB - BA$ is the commutator of the operators A and B. Equation (2.22) is equivalent to

$$\rho(t) = e^{-iHt/\hbar}\rho e^{iHt/\hbar},$$

(2.23)

with $\rho = \rho(0)$, the operator state at the initial time $t = 0$.

Formulas (2.21) and (2.22) should not be confused with the equation given the evolution of *observables* with respect to time. If A is an operator representing a quantum observable, this equation can be obtained by using the following *duality relation*:

$$\mathrm{Tr}\,\rho(t)\,A = \mathrm{Tr}\,e^{-iHt/\hbar}\rho\,e^{iHt/\hbar}\,A = \mathrm{Tr}\,\rho\,e^{iHt/\hbar}\,A\,e^{-iHt/\hbar} = \mathrm{Tr}\,\rho\,A(t)\,,\,(2.24)$$

where we have made use of the invariance of the trace under a cyclic permutation of the involved operators, i.e., $\mathrm{Tr}\,[ABC] = \mathrm{Tr}\,[CAB] = \mathrm{Tr}\,[BCA]$. The symbol Tr means *trace*. In order to obtain the trace of an operator A, we take any orthonormal basis $\{|n\rangle\}$ in the Hilbert space $\mathcal{H}$ and define the trace of A, $\mathrm{Tr}\,A$, by the following series of complex numbers:

$$\mathrm{Tr}\,A := \sum_n \langle n|A|n \rangle.$$

(2.25)

The trace is independent on the orthonormal basis chosen. In a Hilbert space of infinite dimension, like the ones that we usually consider in quantum statistical mechanics, an operator A may have either finite trace, infinite trace or undefined (not defined) trace. Operators representing quantum states have trace one, the identity I ($I\psi = \psi$) has infinite trace and Hamiltonians have undefined trace in general.

In (2.24), we have defined the evolution $A(t)$ of the observable $A = A(t = 0)$ as

$$A(t) = e^{iHt/\hbar} A e^{-iHt/\hbar}. \tag{2.26}$$

This equation is equivalent to

$$\dot{A}(t) = \frac{d}{dt} A(t) = \frac{i}{\hbar} [H, A] \Longleftrightarrow -i\hbar \dot{A}(t) = [H, A]. \tag{2.27}$$

Comparing (2.27) with (2.22), we see that these two equations are similar, although differ in a sign.

The average $\overline{A}$ of the observable represented by the operator A on a given state $\psi = \sum_n c_n |n\rangle$ is given by

$$\overline{A} = \langle \psi | A | \psi \rangle = \sum_{nm} c_n^* c_m \langle n | A | m \rangle. \tag{2.28}$$

In general, the Hilbert space $\mathcal{H}$ is a space of complex functions over some field, so that the scalar product is usually an integral. If $\langle q | n \rangle \equiv \psi_n(q)$, where q is the coordinate, we write

$$\langle n | A | m \rangle = \int \psi_n^*(q) A \psi_m(q) dq, \tag{2.29}$$

where the star denotes complex conjugation.

At this point, it seems convenient to give a precise definition of operator state, i.e., an operator which defines a quantum state. This is an operator ρ on $\mathcal{H}$ with the following properties:

1. Self-adjointness, $\rho^\dagger = \rho$, where the dagger denotes operator adjoint. Thus, any operator state may also be interpreted as a quantum observable.

2. Positivity: for any vector ψ in the Hilbert state $\mathcal{H}$, we have that $\langle \psi | \rho | \psi \rangle \geq 0$.

3. Trace one, Tr $\rho = 1$.

We shall discuss later in the chapter that any operator state ρ is *diagonal* in some orthonormal basis. This means that there exists a basis $|n\rangle$ in $\mathcal{H}$ such that

$$\rho = \sum_n \omega_n |n\rangle \langle n|, \tag{2.30}$$

where $\{\omega_n\}$ are, in principle, complex numbers. Apply ρ to any vector $|m\rangle$ in the above orthonormal basis, so as to obtain

$$\rho |m\rangle = \sum_n \omega_n |n\rangle \langle n | m \rangle = \sum_n \omega_n |n\rangle \delta_{nm} = \omega_m |m\rangle. \tag{2.31}$$

Therefore, ω_m are the eigenvalues of ρ with corresponding eigenvectors $|m\rangle$, $\rho|m\rangle = \omega_m|m\rangle$. The positivity of ρ implies that the numbers ω_m are all non-negative ($\omega_m \geq 0$):

$$0 \leq \langle n|\rho|n\rangle = \omega_n\langle n|n\rangle = \omega_n. \tag{2.32}$$

By definition, the trace of ρ is one. We can calculate the trace using the basis $\{|n\rangle\}$, so that

$$1 = \text{Tr}\,\rho = \sum_n \langle n|\rho|n\rangle = \sum_n \langle n| \left[\sum_m \omega_m |m\rangle\langle m| \right] |n\rangle$$

$$= \sum_n \sum_m \omega_m \langle n|m\rangle\langle m|n\rangle = \sum_n \sum_m \omega_m \delta_{nm} = \sum_n \omega_n = 1. \tag{2.33}$$

Since the numbers ω_n are non-negative, i.e., either zero or positive, equation (2.33) suggests that they represent some sort of probability linked to the number n. In fact, the usual interpretation of a mixed state or a mixture says that the system in the state ρ is in some of the pure states represented by the basis vectors $\{|n\rangle\}$, in which one, we do not know, but the probability that the state be just $|n\rangle$ is ω_n.

Note that a pure state represented by the vector ψ can be also written in the form of state operator as the dyadic product $\rho = |\psi\rangle\langle\psi|$. We recall that any vector representing a state has norm one. We can always construct an orthonormal basis $\{|n\rangle\}$ such that $\psi = |1\rangle$. Therefore, this is the basis that diagonalizes the operator $\rho = |\psi\rangle\langle\psi|$.

This interpretation of a quantum mixed state, along the definition (2.28) of the average of an observable A on a pure state ψ, suggests the definition of the average of A in a mixed state ρ as

$$\overline{A} = \sum_n \omega_n \langle n|A|n\rangle = \text{Tr}\,(\rho\,A). \tag{2.34}$$

In real experimental situations, averages are the real data, which is known in a particular situation. No wonder that the connection between time evolution of states (Schrödinger representation) and observables (Heisenberg representation) given by (2.24) makes use of averages.

2.1.3.1 Connection to the Statistic Formalism

For this connection, the starting point is the definition of the Gibbs probability as

$$p_n = \frac{e^{-\beta E_n}}{Z}. \tag{2.35}$$

Let us identify the numbers p_n, for all n, with the ω_n above considered, so that $p_n = \omega_n$ for all n. Then, we may construct a state operator ρ, often called *density operator* in statistical mechanics, as

$$\rho := \sum_n \frac{1}{Z} e^{-\beta E_n} |n\rangle\langle n| = \frac{1}{Z} e^{-\beta H} \left(\sum_n |n\rangle\langle n| \right). \tag{2.36}$$

Let us apply the expression between parenthesis in (2.36) to an arbitrary vector ψ. Since the vectors $\{|n\rangle\}$ form an orthonormal basis, we have that

$$\left(\sum_n |n\rangle\langle n| \right) |\psi\rangle = \sum_n |n\rangle\langle n|\psi\rangle = \psi = I\psi. \tag{2.37}$$

Therefore, this sum is the identity operator I. Consequently, (2.36) is

$$\rho = \frac{e^{-\beta H}}{\mathrm{Tr}\,(e^{-\beta H})}. \tag{2.38}$$

If we consider the non-normalized state operator $\hat{\rho} := e^{-\beta H}$ (note that $\mathrm{Tr}\,\hat{\rho} = Z$) and derive $\hat{\rho}$ with respect to β, we have

$$\frac{\partial \hat{\rho}}{\partial \beta} = -H\hat{\rho}, \tag{2.39}$$

so that

$$\left(\frac{\partial}{\partial \beta} + H \right) \hat{\rho} = 0. \tag{2.40}$$

It is interesting to compare this equation with the Schrödinger equation , by making the change $\frac{\partial}{\partial \beta} \longmapsto \frac{\hbar}{i} \frac{\partial}{\partial t}$. In any case, if we take the derivative with respect to β of the *normalized* state vector ρ (2.38), it results in

$$\frac{\partial}{\partial \beta} \rho = \frac{\partial}{\partial \beta} \left(\frac{e^{-\beta H}}{Z} \right) = \frac{(-H\,e^{-\beta H})Z - e^{-\beta H}\,(\partial Z/\partial \beta)}{Z^2}, \tag{2.41}$$

where

$$\frac{\partial Z}{\partial \beta} = \frac{\partial}{\partial \beta} \sum_n \langle n|e^{-\beta H}|n\rangle = \frac{\partial}{\partial \beta} \sum_n e^{-\beta E_n} = -\sum_n E_n\, e^{-\beta E_n}. \tag{2.42}$$

Therefore, (2.41) becomes

$$\frac{\partial}{\partial \beta} \rho = -H \frac{e^{-\beta H}}{Z} + \frac{e^{-\beta H}}{Z} \frac{\sum_n E_n\, e^{-\beta E_n}}{Z}. \tag{2.43}$$

Note that the last factor in (2.43) is the average value of the energy, $\overline{E}$, of a system in the state ρ, then

$$\left(\frac{\partial}{\partial \beta} + H\right) \rho = \overline{E}\,\rho\,, \tag{2.44}$$

or

$$\left(\frac{\partial}{\partial \beta} + H - \overline{E}\right) \rho = 0\,. \tag{2.45}$$

2.2 Averages and Fluctuations

It is interesting to note that, if $\rho = \rho(t)$, then it is also an operator state for any value of time t. As such $\rho(t)$ is characterized by (i.) self-adjointness, (ii.) positivity and (iii.) normalization. We see that these conditions are fulfilled:

(i.) The adjoint operator of $\rho(t)$ is $\rho(t)$.

$$[\rho(t)]^\dagger = \left[e^{-iHt/\hbar}\,\rho(0)\,e^{iHt/\hbar}\right]^\dagger = e^{-iHt/\hbar}\,\rho(0)\,e^{iHt/\hbar} = \rho(t)\,. \tag{2.46}$$

Here, we have applied that $(AB)^\dagger = B^\dagger A^\dagger$, $[e^{-iHt/\hbar}]^\dagger = e^{iHt/\hbar}$ and that $\rho = \rho(0)$ is self-adjoint by hypothesis. Therefore, $\rho(t)$ is self-adjoint for any value of t.

(ii.) For any vector ψ in the Hilbert space $\mathcal{H}$, we have

$$\langle \psi | \rho(t) | \psi \rangle = \langle \psi | e^{-iHt/\hbar}\,\rho(0)\,e^{iHt/\hbar} | \psi \rangle = \langle e^{-iHt/\hbar}\psi | \rho(0) | e^{iHt/\hbar}\psi \rangle = \langle \psi(t) | \rho(0) | \psi(t) \rangle \geq 0\,, \tag{2.47}$$

since $\rho = \rho(0)$ is a positive operator. This shows that $\rho(t)$ is a positive operator.

(iii.) Recall that the trace is invariant under a circular permutation of its arguments. Thus,

$$\mathrm{Tr}\,\rho(t) = \mathrm{Tr}\left[e^{-iHt/\hbar}\,\rho(0)\,e^{iHt/\hbar}\right] = \mathrm{Tr}\left[\rho(0)\,e^{iHt/\hbar}\,e^{-iHt/\hbar}\right] = \mathrm{Tr}\,\rho(0) = 1\,. \tag{2.48}$$

The conclusion is that if $\rho = \rho(0)$ is state operator, so is $\rho(t)$ for any value of time t.

The average or mean value of an observable, represented by some operator A, on the state ρ is given by $\mathrm{Tr}\,[A\,\rho] = \mathrm{Tr}\,[\rho\,A]$. If we use a basis of eigenvalues of the Hamiltonian and use the notation $\rho_{nm} = \langle n | \rho | m \rangle$, $A_{nm} = \langle n | A | m \rangle$, we have

$$\overline{E} = \langle H \rangle = \mathrm{Tr}\,[\rho H] = \sum_n \langle n | \rho H | n \rangle = \sum_{nm} \langle n | \rho | m \rangle \langle m | H | n \rangle$$

$$= \sum_{nm} \langle n | \rho | m \rangle\, E_n \langle n | m \rangle = \sum_{nm} \rho_{nm}\, E_n\, \delta_{nm} = \sum_n E_n\, \rho_{nn}\,. \tag{2.49}$$

The entropy of some quantum state ρ is usually defined as the trace of the operator $\mathfrak{S} = -k\rho \ln \rho$. This operator is well defined, if ρ admits the spectral decomposition $\rho = \sum_n \lambda_n |n\rangle\langle n|$, where $\{|n\rangle\}$ is some basis in the Hilbert space $\mathcal{H}$, we take

$$\ln \rho := \sum_n (\ln \lambda_n)|n\rangle\langle n|, \qquad (2.50)$$

where we have omitted in the sum in (2.50) those terms for which $\lambda_m = 0$ (and therefore do not appear in the spectral decomposition for ρ). This spectral decomposition, being ρ a self adjoint tracial operator, always exists.

Note that

1. Since $\lambda_n \leq 1, \ln \lambda_n \leq 0$ and therefore, we need the minus sign in the definition of the entropy operator in order to obtain a positive expression.

2. If ρ represents a pure state, then there exists a vector $|1\rangle$ so that $\rho = |1\rangle\langle 1|$. Then, $\lambda_1 = 1$ and $\lambda_n = 0$ for $n \neq 1$. After the previous definition for the logarithm, $\ln \rho = 0$. This means that the entropy for a pure state is always zero.

As a matter of fact, the entropy is a magnitude, and hence it should be a number. Then, we define the entropy for ρ into two equivalent forms:

(i.) The entropy is the trace of the entropy operator $\mathfrak{S} = -k_B \rho \ln \rho$:

$$S := \mathrm{Tr}\, \mathfrak{S} = -k_B \sum_n \langle n|\rho \ln \rho|n\rangle = -k_B \sum_{nm} \langle n|\rho|m\rangle\langle n|\ln \rho|n\rangle = -k_B \sum_{nm} \rho_{nm} \langle n|\ln \rho|n\rangle. \quad (2.51)$$

(ii.) From (2.51), we see that S is also the average on ρ of $\ln \rho$:

$$S := -k_B \langle \overline{\ln \rho}\rangle = -k_B \mathrm{Tr}\,[\rho \ln \rho] = -k_B \sum_n \langle n|\rho \ln \rho|n\rangle. \quad (2.52)$$

Our goal is to calculate the entropy for the canonical state $\rho = e^{-\beta H}/(\mathrm{Tr}\, e^{-\beta H})$. First, note that for this state

$$\ln \rho = -\beta H - \ln\{\mathrm{Tr}\,[e^{-\beta H}]\}. \quad (2.53)$$

Then using the expression (2.51), the form of the entropy reads

$$S = -k_B \sum_{nm} \rho_{nm}[-\beta\langle m|H|n\rangle - \langle m|\ln\{\mathrm{Tr}\,(e^{-\beta H})\}|n\rangle]$$

$$= k_B\beta \sum_{nm} \rho_{nm} \langle m|H|n\rangle + k_B[\ln\{\mathrm{Tr}\,(e^{-\beta H})\}] \sum_{nm} \rho_{nm} \langle m|n\rangle. \quad (2.54)$$

Now, assuming that $\{|n\rangle\}$ is the basis of eigenvectors of the Hamiltonian H and taking into account that $\langle m|n\rangle = \delta_{nm}$ and $\sum_n \rho_{nn} = \sum_n \langle n|\rho|n\rangle = 1$, equation (2.54) becomes

$$S = k_B\beta \sum_n \rho_{nn} E_n + k_B \ln\{\mathrm{Tr}\,(e^{-\beta H})\} = k_B\overline{E} + k_B \ln\{\mathrm{Tr}\,(e^{-\beta H})\}. \quad (2.55)$$

Thus, taking into account the thermodynamic relation free energy with entropy,

$$F = \overline{E} - \frac{1}{k_B \beta} S \iff S = k_B \beta \overline{E} + k_B[-\beta F], \qquad (2.56)$$

we conclude that the free energy for the canonical state is

$$F = -\frac{1}{\beta} \ln(\mathrm{Tr}\, e^{-\beta H}). \qquad (2.57)$$

We may write S in another form. Take (2.50) and multiply it by $-k_B \rho = -k_B \sum_m \lambda_m |m\rangle\langle m|$, taking into account that $\langle n|m\rangle = \delta_{nm}$. This gives

$$\mathfrak{S} = -k_B \sum_n \lambda_n (\ln \lambda_n)|n\rangle\langle n| \implies S = -k_B \sum_n \lambda_n \ln \lambda_n. \qquad (2.58)$$

Remark The analogous classical expressions are given in terms of probabilities or, equivalently, in terms of average occupation numbers. Taken the continuous limit, we get

$$F = -k_B T \ln \int e^{-E(p,q)/k_B T} \, dp \, dq \qquad (2.59)$$

for the free energy, where $E(p, q)$ is the energy in terms of positions and momenta and T the temperature, and

$$\rho(p,q) = (2\pi\hbar)^{-s} e^{(F-E(p,q))/k_B T}, \qquad (2.60)$$

for the density (state), where s is the number of degrees of freedom.

2.3 Summary

- The canonical equilibrium state is

$$\rho = \frac{1}{\mathrm{Tr}\, e^{-\beta H}} e^{-\beta H}.$$

- The free energy is

$$F = -\frac{1}{\beta} \ln \mathrm{Tr}\, e^{-\beta H}.$$

- From the above, the canonical state is

$$\rho = e^{\beta(F-H)} .$$

- The entropy is given by the quantum counterpart of the Boltzmann definition

$$S = -k_B \operatorname{Tr}(\rho \ln \rho) = k_B \, \beta \operatorname{Tr}((H-F)\rho) .$$

- For each state operator ρ, there exists a basis $\{|n\rangle\}$ in the Hilbert space such that

$$\rho = \sum_n \lambda_n |n\rangle\langle n| ,$$

where $\lambda_n \geq 0$. Then, the entropy in terms of the λ_n is

$$S = -k_B \sum_n \lambda_n \ln \lambda_n .$$

Since $\lambda_n \leq 1$, then $\ln \lambda_n \leq 0$ and hence, S is always positive definite.
- Assume that there exists a basis $\{|n\rangle\}$ of eigenvectors of the Hamiltonian, $H|n\rangle = E_n|n\rangle$. Since the canonical state ρ is a function on the Hamiltonian H, both operators commute, $[H, \rho] = 0$ and therefore have the same basis of eigenvectors. Now,

$$\lambda_n = \frac{e^{-\beta E_n}}{\sum_n e^{-\beta E_n}} .$$

2.3.1 H-Theorem and Approach to Equilibrium

Let us consider an arbitrary isolated system that may be in different states. Let us call $P_r(t)$ the probability that the system be in the state r at time t. Let us assume the existence of transitions between states. Let us call $W_{r \to s}$ to the transition probability per unit of time between the state r to the state s. Then, for all value of t, we have that

$$\frac{dP_r}{dt} = \sum_s P_s\, W_{s \to r} - \sum_s P_r\, W_{r \to s} . \tag{2.61}$$

It is often reasonable to assume that $W_{s \to r} = W_{r \to s}$. In this case, we write $W_{rs} := W_{s \to r}$ for simplicity. Then, equation (2.61) has the following form:

$$\frac{dP_r}{dt} = \sum_s W_{rs}(P_s - P_r) . \tag{2.62}$$

Defining the following magnitude [2]

$$\mathbb{H} := \sum_r P_r \ln P_r \,, \tag{2.63}$$

where the index r runs out all possible states of the system. If we derive $\mathbb{H}$ with respect to time, one gets

$$\frac{d\mathbb{H}}{dt} = \sum_r \left\{ \frac{dP_r}{dt} \ln P_r + P_r \frac{1}{P_r} \frac{dP_r}{dt} \right\} = \sum_r \left\{ (1 + \ln P_r) \frac{dP_r}{dt} \right\} \,. \tag{2.64}$$

Using (2.63) in (2.64), to have

$$\frac{d\mathbb{H}}{dt} = \sum_r \left(\sum_s W_{rs}(P_s - P_r)(1 + \ln P_r) \right)$$

$$= \frac{1}{2} \sum_{rs} (W_{rs}(P_s - P_r)(1 + \ln P_r) + W_{rs}(P_r - P_s)(1 + \ln P_s))$$

$$= \frac{1}{2} \sum_{rs} (P_s \ln P_r + P_s - P_r \ln P_r - P_r + P_r \ln P_s + P_r - P_s \ln P_s - P_s)$$

$$= \frac{1}{2} \sum_{rs} (W_{rs}[P_r(\ln P_s - \ln P_r)] + W_{rs}[-P_s(\ln P_s - \ln P_r)])$$

$$= -\frac{1}{2} \sum_{rs} W_{rs}(P_s - P_r)(\ln P_s - \ln P_r) \,. \tag{2.65}$$

Then, if

$$P_s > P_r \implies \ln P_s > \ln P_r \implies (P_s - P_r)(\ln P_s - \ln P_r) > 0 \,, \tag{2.66}$$
$$P_s < P_r \implies \ln P_s < \ln P_r \implies (P_s - P_r)(\ln P_s - \ln P_r) > 0 \,, \tag{2.67}$$
$$P_s = P_r \implies \ln P_s = \ln P_r \implies (P_s - P_r)(\ln P_s - \ln P_r) = 0 \,. \tag{2.68}$$

These formulas imply that

$$\boxed{\frac{d\mathbb{H}}{dt} \leq 0} \,. \tag{2.69}$$

This result is the well-known *Boltzmann H-theorem*. The variation with time of the magnitude $\mathbb{H}$ is either zero or negative. According to (2.64), this variation is zero if and only if $P_r(t)$ is constant with time, i.e., $dP_r(t)/dt = 0$. If $P_r(t)$ varies with time, then

$$\frac{d\mathbb{H}}{dt} < 0 \,. \tag{2.70}$$

Then, since $S = -k_B \mathbb{H}$, we have that

$$\frac{dS}{dt} \geq 0, \tag{2.71}$$

or the entropy of a statistic system is always *monotonically increasing*. Note that this result is equally valid for both classical and quantum systems [12,13].

2.4   Perturbative Method for the Density Operator

The non-normalized statistical operator $\hat{\rho} = e^{-\beta H}$ satisfies equation (2.40), which is

$$\frac{\partial \hat{\rho}}{\partial \beta} = -H\hat{\rho}. \tag{2.72}$$

This equation gives us the variation of the (non-normalized) statistical operator with the temperature. Often, $H = H_0 + \lambda V$, where H_0 is chosen under the condition than an equation of the form:

$$\frac{\partial \rho_0}{\partial \beta} = -H_0 \rho_0, \tag{2.73}$$

must have a solution. Here V is some *potential* and λ a real number. As solution of (2.73), we choose $\rho_0(\beta) = e^{-\beta H_0}$ and assume that $\rho_0(\beta)$ has been determined. If $\rho = e^{-\beta H}$ is the solution of (2.72) that we want to determine, we may write

$$\frac{\partial}{\partial \beta}(e^{\beta H_0} \rho) = H_0 e^{\beta H_0} \rho + e^{\beta H_0} \frac{\partial \rho}{\partial \beta} = e^{\beta H_0} H_0 \rho + e^{\beta H_0}(-H\rho)$$

$$= e^{\beta H_0}(H_0 - H)\rho = e^{\beta H_0}(-\lambda V)\rho = -\lambda e^{\beta H_0} V \rho. \tag{2.74}$$

Let us integrate this identity between the integration limits 0 and β. It gives

$$[e^{\beta H_0} \rho]_0^\beta = -\lambda \int_0^\beta e^{\beta' H_0} V \rho(\beta') d\beta', \tag{2.75}$$

which with the initial condition $\rho(0) = I$, the identity operator,[1] we have

$$\rho(\beta) = e^{-\beta H_0} \left[I - \lambda \int_0^\beta e^{\beta' H_0} V \rho(\beta') d\beta' \right]$$

$$= \rho_0(\beta) \left[I - \lambda \int_0^\beta e^{\beta' H_0} V \rho(\beta') d\beta' \right] = \rho_0(\beta) - \lambda \int_0^\beta \rho_0(\beta - \beta') V \rho(\beta') d\beta'. \tag{2.76}$$

---

[1] The identity operator cannot be used as an operator state, since it has infinite trace.

For convenience, if we use the following notation,

$$K_0(\beta - \beta') := \rho_0(\beta - \beta') V, \tag{2.77}$$

then (2.76) takes the form

$$\rho(\beta) = \rho_0(\beta) - \lambda \int_0^\beta K_0(\beta - \beta') \rho(\beta') d\beta'. \tag{2.78}$$

2.4.1 The Partition Function

Introducing the following operators,

$$U(\tau, \tau') := e^{\tau H_0} e^{-(\tau - \tau')H} e^{-\tau' H_0} \tag{2.79}$$

and

$$V(\tau) = e^{\tau H_0} V e^{-\tau H_0}, \tag{2.80}$$

we have

$$U(\beta, 0) = e^{\beta H_0} e^{-\beta H} \iff e^{-\beta H} = e^{-\beta H_0} U(\beta, 0). \tag{2.81}$$

Thus, we may write the partition function corresponding to H as

$$Z = \text{Tr}[e^{-\beta H}] = \text{Tr}[e^{-\beta H_0} U(\beta, 0)]. \tag{2.82}$$

By taking partial derivative in (2.79) with respect to the variable τ,

$$\frac{\partial U}{\partial \tau} = e^{\tau H_0} H_0 e^{-(\tau - \tau')H} e^{-\tau' H_0} + e^{\tau H_0}(-H) e^{-(\tau - \tau')H} e^{-\tau' H_0}$$

$$-\lambda \left(e^{\tau H_0} V e^{-\tau H_0} \right) \left(e^{\tau H_0} e^{-(\tau - \tau')H} e^{-\tau' H_0} \right) = -\lambda V(\tau) U(\tau, \tau'). \tag{2.83}$$

Performing the integration of (2.83) with limits τ' and τ, taking into account that $U(\tau, \tau) = I$ for all τ yields

$$U(\tau, \tau') = I - \lambda \int_{\tau'}^\tau V(\tau_1) U(\tau_1, \tau') d\tau_1. \tag{2.84}$$

This is an integral equation that may be solved by using an iterative method. This is done by replacing $U(\tau_1, \tau')$ under the integral sign in (2.84) by its expression given in equation (2.84). It results in a double integral under which appears the term $U(\tau_2, \tau')$. We repeat this procedure indefinitely, so as to obtain the following expression:

$$U(\tau, \tau') = \sum_{n=0}^{\infty} (-\lambda)^n \int_{\tau'}^{\tau} \int_{\tau'}^{\tau_1} \cdots \int_{\tau'}^{\tau_{n-1}} V(\tau_1) \dots V(\tau_n) \, d\tau_1 \dots d\tau_n \, . \quad (2.85)$$

This gives for the partition function

$$Z = \text{Tr}\,[e^{-\beta H}] = \text{Tr}\,[e^{-\beta H_0}\, U(\beta, 0)]$$

$$= \sum_{n=0}^{\infty} (-\lambda)^n \int_0^{\beta} \int_0^{\beta_1} \cdots \int_0^{\beta_{n-1}} \text{Tr}\,[e^{-\beta H_0}\, V(\beta_1) \dots V(\beta_n)] \, d\beta_1 \dots d\beta_n \, . \quad (2.86)$$

Observe that in (2.86), $\tau' = 0$, $\tau = \beta$, $\tau_k = \beta_k$. Also $e^{-\beta H_0}$ can be inserted inside the integral because there is not an integration over the variable β. This gives the partition function for H in terms of a series in known quantities like $e^{\tau H_0}$, which is known by hypothesis, and the potential V.

The multiple integral in (2.86) can be written as

$$\int_0^{\beta} \int_0^{\beta_1} \cdots \int_0^{\beta_{n-1}} \text{Tr}\,[e^{-\beta H_0}\, V(\beta_1) \dots V(\beta_n)] \, d\beta_1 \dots d\beta_n$$

$$= \frac{1}{n!} \int_0^{\beta} \cdots \int_0^{\beta} \text{Tr}\left[e^{-\beta H_0}\, V(\tau_1) \dots V(\tau_n)\right] d\tau_1 \dots d\tau_n \quad (2.87)$$

with $\tau_1 > \tau_2 > \cdots > \tau_n$. Therefore, we have an ordered product $V(\tau_1) \dots V(\tau_n)$ in terms of the arguments τ_i. This ordered product goes from the lowest temperature (highest τ) to the highest temperature (lowest τ). From this point of view, the partition function (2.86) can be written as

$$Z = \text{Tr}\,[e^{-\beta H_0}] \sum_{n=0}^{\infty} \lambda^n \, W(n) \, , \quad (2.88)$$

where

$$W(n) = [\text{Tr}\, e^{-\beta H_0}]^{-1} \frac{(-1)^n}{n!} \int_0^{\beta} \cdots \int_0^{\beta} \text{Tr}\left[e^{-\beta H_0}\, V(\tau_1) \dots V(\tau_n)\right] d\tau_1 \dots d\tau_n \, . \quad (2.89)$$

Another possible form for the partition function is

$$Z = \text{Tr}\,[e^{-\beta H_0}] \exp\left\{\sum_{l=1}^{\infty} \lambda^l \, U_l\right\} \, , \quad (2.90)$$

where the U_l are related to the $W(n)$. This relation can be obtained through the identity between (2.88) and (2.90):

$$\exp\left\{\sum_{l=1}^{\infty} \lambda^l U_l\right\} = 1 + \sum_{l=1}^{\infty} \lambda^l U_l + \frac{1}{2!} \sum_{ll'} \lambda^{l+l'} U_l U_{l'} + \dots$$

$$= 1 + \lambda W(1) + \lambda^2 W(2) + \dots , \qquad (2.91)$$

which gives some relations between U_l and $W(n)$ such as

$$W(1) = U_1, \qquad W(2) = U_2 + \frac{1}{2} U_2^2 , \dots , \qquad (2.92)$$

etc. This exponential form is quite useful to determine the structure of the thermodynamic potential (grand canonical free energy) Ω. Since

$$Z = e^{-\beta \Omega} \implies \Omega = -\frac{1}{\beta} \ln Z , \qquad (2.93)$$

we have

$$\Omega = -\frac{1}{\beta} \ln \left[\mathrm{Tr}\,(e^{-\beta H_0}) \exp\left\{ \sum_{l=1}^{\infty} \lambda^l U_l \right\} \right] \qquad (2.94)$$

and, hence,

$$\Omega = \frac{1}{\beta} \ln \left\{ \mathrm{Tr}\,(e^{-\beta H_0}) \right\} - \frac{1}{\beta} \sum_{l=1}^{\infty} \lambda^l U_l . \qquad (2.95)$$

2.5   Summary

- In this chapter, we have re-obtained the expressions for the statistical averages introduced earlier. In doing so, we have established the equivalences with mean values of dynamical quantities, starting from the density operator ρ (2.40) and the free energy F, (2.57).
- The density operator ρ was expressed in terms of the cluster expansion to accommodate interactions, thus allowing for a perturbative expansion.
- By enunciating the Boltzmann H-theorem , we are able to determine a direction of the time evolution of the entropy.

Operator Representations of the Statistical Mechanics

3

3.1 Introduction

As is common in quantum field theory, we shall introduce creation and annihilation operators for particles, either bosons or fermions, and their corresponding commutation or anticommutation relations. The time dependence of the operators is that governed by the Hamiltonian density. The passage to the statistical mechanics framework requires the addition, to the time variable, of a *temperature-like* variable. This can be done with the help of time-temperature representations, or, equivalently, by using a complex time variable. Other possibility to deal with the statistical mechanics in terms of operators and equations of motion is the use of dual theories of which the thermal field theory is an example, as well as the so-called closed-path integration [14,15].

In this chapter, we are presenting the notions of thermal propagators, thermal ordering and thermal Green's functions. From there, we shall recover the statistical averages, presented in the previous chapters, this time in terms of amplitudes rather than probabilities. An important point is the inclusion of interactions, an ingredient relevant to the treatment of real physical systems.

We assume that the reader is already familiar with the formalism of second quantization. Nevertheless, the material of this chapter is self-contained. In the discussion of double (time-temperature) Green's functions, we shall follow closely the presentation given in his book by Ter Haar [16].

O. Civitarese and M. Gadella, *Methods in Statistical Mechanics*, Lecture Notes
in Physics 974, https://doi.org/10.1007/978-3-030-53658-9_3

3.2 The Concept of Thermal Propagator

To begin with, let us list some concepts to be used in this section:

- *Statistical averages*. If A is an operator which represents a quantum observable, its statistical average is defined as

$$\langle A \rangle := \frac{\text{Tr}\,(e^{-\beta H}\,A)}{\text{Tr}\,e^{-\beta H}}\,, \tag{3.1}$$

where H is the Hamiltonian which governs the time evolution of the quantum system.

- *Thermal dependence*. Let a_k and $a_k^{\dagger}$ be the annihilation and creation operators for a particle with momentum k. For a given parameter τ, we define

$$a_k(\tau) = e^{\tau H}\,a_k\,e^{-\tau H}\,, \tag{3.2}$$

$$a_k^{\dagger}(\tau) = e^{\tau H}\,a_k^{\dagger}\,e^{-\tau H}\,. \tag{3.3}$$

The denomination of thermal dependence comes after the identification of τ with a magnitude proportional to the inverse of the temperature. Note that τ is playing the role of an imaginary time, since the time evolution of the operators is controlled by e^{-itH}; then, $\tau \Leftrightarrow it$.

- *Thermal ordering*. Let $A(\tau)$ and $B(\tau)$ be two operators depending on the variable τ. Then, the thermal ordering of $A(\tau_1)$ and $B(\tau_2)$, $T[A(\tau_1)B(\tau_2)]$, is defined as

$$T_\tau[A(\tau_1)B(\tau_2)] := A(\tau_1)B(\tau_2)\,, \quad \text{if } \tau_1 > \tau_2\,,$$

$$T_\tau[A(\tau_1)B(\tau_2)] := \pm B(\tau_2)A(\tau_1)\,, \quad \text{if } \tau_1 < \tau_2\,. \tag{3.4}$$

In the second row of (3.4), the sign plus corresponds to bosons, while the sign minus corresponds to fermions.

Next, let us introduce the definition of the *thermal propagator* for a system of particles; it is given by

$$\boxed{G(k, k', \tau - \tau') := -\langle T_\tau[a_k(\tau)\,a_{k'}^{\dagger}(\tau')]\rangle}\,. \tag{3.5}$$

In analogy with similar expressions that appear in other branches of physics, we shall refer the thermal operator as *Green's function*.

We begin our analysis with the simplest case, which is that of free particles. We shall find Green's function for Hamiltonians that can be written in diagonal form in some given basis. Take H_0 as an *unperturbed* Hamiltonian, which is written in terms of creation and annihilation operators as

$$H_0 := \sum_k \mathcal{E}_k\,a_k^{\dagger}\,a_k\,, \tag{3.6}$$

where k is a finite or infinite sequence of natural numbers, labeling the energy levels $\mathcal{E}_k$. The thermal evolution of the operators $a_k^\dagger$ and a_k is governed by H_0:

$$a_k^\dagger(\tau) := e^{\tau H_0} a_k^\dagger e^{-\tau H_0} . \tag{3.7}$$

Our first task is to calculate the commutator,

$$[H_0, a_k^\dagger(\tau)] = [H_0, e^{\tau H_0} a_k^\dagger e^{-\tau H_0}] = H_0 e^{\tau H_0} a_k^\dagger e^{-\tau H_0} - e^{\tau H_0} a_k^\dagger e^{-\tau H_0} H_0$$
$$= e^{\tau H_0}(H_0 a_k^\dagger - a_k^\dagger H_0)e^{-\tau H_0} = e^{\tau H_0}[H_0, a_k^\dagger]e^{-\tau H_0} , \tag{3.8}$$

where we have used the fact that H_0 commutes with any function of it. Observe that

$$\frac{\partial}{\partial \tau}[a_k^\dagger(\tau)] = H_0 e^{\tau H_0} a_k^\dagger e^{-\tau H_0} - e^{\tau H_0} a_k^\dagger e^{-\tau H_0} H_0 . \tag{3.9}$$

Combining (3.8) and (3.9), we find that

$$\frac{\partial}{\partial \tau}[a_k^\dagger(\tau)] = [H_0, a_k^\dagger(\tau)] = e^{\tau H_0}[H_0, a_k^\dagger]e^{-\tau H_0} . \tag{3.10}$$

The final expression for the commutator $[H_0, a_k^\dagger]$ will depend on the nature of particles, which may be either fermions or bosons. When the set of energy levels is discrete, creation and annihilation operators satisfy the following relations:

$$[a_n, a_m^\dagger] = a_n a_m^\dagger - a_m^\dagger a_n = \delta_{nm} , \quad \text{for bosons} \tag{3.11}$$
$$\{a_n, a_m^\dagger\} = a_n a_m^\dagger + a_m^\dagger a_n = \delta_{nm} , \quad \text{for fermions.} \tag{3.12}$$

All other commutators or anticommutators vanish. The symbol δ_{nm} is the Kronecker delta. From the above equations, we readily obtain the following relations:

$$a_m a_n^\dagger = [a_m, a_n^\dagger] + a_n^\dagger a_m = \delta_{mn} + a_n^\dagger a_m , \quad \text{for bosons} \tag{3.13}$$
$$a_m a_n^\dagger = \{a_m, a_n^\dagger\} - a_n^\dagger a_m = \delta_{mn} - a_n^\dagger a_m , \quad \text{for fermions.} \tag{3.14}$$

Consequently, for bosons we have that

$$H_0 a_k^\dagger = \sum_n \mathcal{E}_n a_n^\dagger a_n a_k^\dagger = \sum_n \mathcal{E}_n a_n^\dagger([a_n, a_k^\dagger] + a_k^\dagger a_n)$$
$$= \sum_n \mathcal{E}_n a_n^\dagger(\delta_{nk} + a_k^\dagger a_n) = \mathcal{E}_k a_k^\dagger + a_k^\dagger H_0 , \tag{3.15}$$

and for fermions

$$H_0 a_k^\dagger = \sum_n \mathcal{E}_n a_n^\dagger a_n a_k^\dagger = \sum_n \mathcal{E}_n a_n^\dagger(\{a_n, a_k^\dagger\} - a_k^\dagger a_n)$$
$$= \sum_n \mathcal{E}_n a_n^\dagger(\delta_{nk} - a_k^\dagger a_n) = \mathcal{E}_k a_k^\dagger + a_k^\dagger H_0 . \tag{3.16}$$

In consequence,

$$[H_0, a_k^\dagger] = \mathcal{E}_k a_k^\dagger + a_k^\dagger H_0 - a_k^\dagger H_0 = \mathcal{E}_k a_k^\dagger , \tag{3.17}$$

a relation valid for both, bosons and fermions. In combination with (3.10), we arrive to the equation which determines the thermal evolution of the creation operators, namely,

$$\frac{\partial}{\partial \tau}[a_k^\dagger(\tau)] = e^{\tau H_0}(\mathcal{E}_k a_k^\dagger)e^{-\tau H_0} = \mathcal{E}_k a_k^\dagger(\tau) . \tag{3.18}$$

Analogously, if instead of (3.7) we take the annihilation operator

$$a_k(\tau) := e^{\tau H_0} a_k e^{-\tau H_0} , \tag{3.19}$$

and following the same steps, we get

$$\frac{\partial}{\partial \tau}[a_k(\tau)] = -\mathcal{E}_k a_k^\dagger(\tau) . \tag{3.20}$$

The formal integration of (3.18) and (3.20) yields

$$a_k^\dagger(\tau) = e^{\tau \mathcal{E}_k} a_k^\dagger(0) = e^{\tau \mathcal{E}_k} a_k^\dagger \tag{3.21}$$

$$a_k(\tau) = e^{-\tau \mathcal{E}_k} a_k(0) = e^{-\tau \mathcal{E}_k} a_k . \tag{3.22}$$

There exists another method to obtain the above formulas. Let us use it to derive (3.21) and similarly (3.22). Consider the commutator

$$[e^{\tau H_0}, a_k^\dagger] = e^{\tau H_0} a_k^\dagger - a_k^\dagger e^{\tau H_0} , \tag{3.23}$$

which gives

$$a_k^\dagger(\tau) = e^{\tau H_0} a_k^\dagger e^{-\tau H_0} = [e^{\tau H_0}, a_k^\dagger]e^{-\tau H_0} + a_k^\dagger . \tag{3.24}$$

Write

$$[e^{\tau H_0}, a_k^\dagger] = [\sum_{n=1}^{\infty} \frac{\tau^n}{n!} H_0^n, a_k^\dagger] = \sum_{n=1}^{\infty} \frac{\tau^n}{n!} [H_0^n, a_k^\dagger] . \tag{3.25}$$

Thus, we have to compute the commutators $[H_0^n, a_k^\dagger]$ for all values of n. Let us proceed step by step. For $n = 1$, we have (3.17). For $n = 2$,

$$\begin{aligned}
[H_0^2, a_k^\dagger] &= H_0 H_0 a_k^\dagger - H_0 a_k^\dagger H_0 + H_0 a_k^\dagger H_0 - a_k^\dagger H_0 H_0 \\
&= H_0[H_0, a_k^\dagger] + [H_0, a_k^\dagger]H_0 = \mathcal{E}_k a_k^\dagger H_0 + \mathcal{E}_k H_0 a_k^\dagger \\
&= 2\mathcal{E}_k a_k^\dagger H_0 - \mathcal{E}_k a_k^\dagger H_0 + \mathcal{E}_k H_0 a_k^\dagger = 2\mathcal{E}_k a_k^\dagger H_0 + \mathcal{E}_k [H_0, a_k^\dagger] \\
&= 2\mathcal{E}_k a_k^\dagger H_0 + \mathcal{E}_k^2 a_k^\dagger .
\end{aligned} \tag{3.26}$$

Using the same strategy, we have

$$[H_0^3, a_k^\dagger] = \mathcal{E}_k^3 a_k^\dagger + 2\mathcal{E}_k^2 a_k^\dagger + 4\mathcal{E}_k a_k^\dagger H_0^2 + \mathcal{E}_k^2 a_k^\dagger H_0 + \mathcal{E}_k^2 a_k^\dagger H_0^2 , \qquad (3.27)$$

and so on. Let us use this result in (3.25). We obtain

$$\sum_{n=1}^{\infty} \frac{\tau^n}{n!} [H_0^n, a_k^\dagger] = a_k^\dagger \left(\mathcal{E}_k \tau + \frac{\tau^2}{2!} \mathcal{E}_k^2 + \frac{\tau^3}{3!} \mathcal{E}_k^3 + \dots \right.$$

$$2\mathcal{E}_k \frac{\tau^2}{2!} H_0 + \mathcal{E}_k \frac{\tau^3}{3!} H_0 + \dots$$

$$\left. 4\mathcal{E}_k \frac{\tau^3}{3!} H_0^2 + 4\mathcal{E}_k^2 \frac{\tau^3}{3!} H_0 + \dots \right)$$

$$= a_k^\dagger \left(1 + \tau \mathcal{E}_k + \tau^2 \frac{\mathcal{E}_k}{2} + \dots - 1 \right) \left(1 + \tau H_0 + \frac{\tau^2}{2} H_0^2 + \dots \right)$$

$$= a_k^\dagger (e^{\tau \mathcal{E}_k} - 1) e^{\tau H_0} . \qquad (3.28)$$

Using (3.28) in (3.24), we finally obtain ($a_k^\dagger \equiv a_k^\dagger(0)$):

$$a_k^\dagger(\tau) = (a_k^\dagger(e^{\tau \mathcal{E}_k} - 1) e^{\tau H_0} + a_k^\dagger e^{\tau H_0}) e^{-\tau H_0}$$

$$= (a_k^\dagger e^{\tau \mathcal{E}_k} e^{\tau H_0} - a_k^\dagger e^{\tau H_0} + a_k^\dagger e^{\tau H_0}) e^{-\tau H_0} = a_k^\dagger e^{\tau \mathcal{E}_k} = a_k^\dagger(0) e^{\tau \mathcal{E}_k} . \qquad (3.29)$$

We postpone advancing further on this analysis to introduce, in the next subsection, a very powerful tool: the Wick theorem [18, 19].

3.2.1 Wick Theorem

Let us consider the product of an arbitrary number of creation and annihilation operators. Normal ordering of any product of creation and annihilation operators is defined as the product of the same operators ordered in such a way that all creation operators are written at the left and all the annihilation operators at the right.

Let A and B be either creation ($a_k^\dagger$) or annihilation (a_k) operators. Then, the *contraction* of A and B is defined as

$$\overset{\frown}{AB} = [A, B]_\pm \mp BA \qquad (3.30)$$

where $: AB :$ denotes the normal ordering of the operators A and B.

We note that contractions of creation or annihilation operators are always zero, due to the commutation and anticommutation relations $[a_k, a_{k'}]_\pm = 0$, or $[a_k^\dagger, a_{k'}^\dagger]_\pm = 0$. This result is generalized for any product of any number of creation and annihilation operators, as demonstrated by Wick :

Theorem (Wick). The product of any number of creation and annihilation operators can be written as the sum of all possible non-vanishing contractions among m pairs of operators, $m = 1, 2, \ldots$, plus the normal order, consisting in writing all the creation operators to the right and all the annihilation operators to the left.

To illustrate the theorem, let us take the case of fermions, where the simplest non-vanishing products are

$$a_l \, a_k^\dagger = [a_l, a_k^\dagger]_+ - a_k^\dagger \, a_l = \delta_{lk} - a_k^\dagger \, a_l = \widehat{lk} - a_k^\dagger \, a_l \,, \tag{3.31}$$

where the symbol $\widehat{lk}$ represents the Kronecker delta $\delta_{l,k}$. If we take a product of one annihilation operator and two creation operators, we write according to previous results:

$$a_l \, a_m^\dagger \, a_k^\dagger = (\widehat{lm} - a_m^\dagger \, a_l) a_k^\dagger = \widehat{lm} \, a_k^\dagger - a_m^\dagger (\widehat{lk} - a_k^\dagger \, a_l) = \widehat{lm} \, a_k^\dagger - a_m^\dagger \, \widehat{lk} + a_m^\dagger \, a_k^\dagger \, a_l \tag{3.32}$$

or

$$a_l \, a_m \, a_k^\dagger = a_l (\widehat{mk} - a_k^\dagger \, a_m) = \widehat{mk} \, a_l - (\widehat{lk} - a_k^\dagger \, a_l) a_m = \widehat{mk} \, a_l - \widehat{lk} \, a_m + a_k^\dagger \, a_l \, a_m \,. \tag{3.33}$$

Then, for four operators like

$$a_l \, a_m \, a_k^\dagger \, a_w^\dagger = \widehat{mk} \, a_l \, a_w^\dagger - \widehat{lk} \, a_m \, a_w^\dagger + a_k^\dagger \, a_l \, a_m \, a_w^\dagger$$
$$= \widehat{mk}(\widehat{lw} - a_w^\dagger \, a_l) - \widehat{lk}(\widehat{mw} - a_w^\dagger \, a_m) + a_k^\dagger \, a_l (\widehat{mw} - a_w^\dagger \, a_m) \,, \tag{3.34}$$

which yields to the following final expression:

$$= (\widehat{mk} \, \widehat{lw} - \widehat{lk} \, \widehat{mw}) + (\widehat{lk} \, a_w^\dagger \, a_m + \widehat{mw} \, a_k^\dagger \, a_l - \widehat{mk} \, a_w^\dagger \, a_l - \widehat{lw} \, a_k^\dagger \, a_m) + (a_k^\dagger \, a_w^\dagger \, a_l \, a_m) \,. \tag{3.35}$$

The first parenthesis in (3.35) represents all possible fully contracted terms, the second one is the sum of all products of one contraction and one pair of ordered operators and the third term is the fully ordered set of the four operators.

Therefore, for any arbitrary number of creation and annihilation operators, we have that

Product $(a_l \, a_m \ldots a_k^\dagger \, a_w^\dagger \ldots) \equiv$ Normal ordered term + sum of 1 contraction $\otimes$ ordered $(N - 2)$ + sum of 2 contractions $\otimes$ ordered $(N - 4)$ + $\ldots\ldots$ + sum of all contractions.

Note that for bosons the contributions of all terms are positive, while for fermions each component is multiplied by a sign.

Thus, this procedure is easily generalized since any product of N operators can be decomposed into cyclic combinations of $1 \otimes (N - 1) + 2 \otimes (N - 2) + \cdots + N$ ordered operators. As we have seen, for the case of fermions the $\pm$ sign in front of the terms with certain number of contractions is given by $(-1)^{N_p}$, N_p being the number of permutation among the operators needed to bring side by side the two operators to be contracted. For bosons all contributions are positive.

Why is this theorem so relevant? Often, we have to calculate averages of products of creation and annihilation operators acting on some reference state, like the vacuum state, a state which is annihilated by any annihilation operator:

$$a_k|0\rangle = 0, \qquad (3.36)$$

or the one particle state, $|k\rangle = a_k^\dagger|0\rangle$, or the N particle state $|k_1, k_2, \ldots, k_N\rangle = \prod_{i=1}^{N} a_i^\dagger|0\rangle$.

Thus, the use of the Wick theorem greatly simplifies the calculation of vacuum expectation values of products of creation and annihilation operators, as well as the calculation of expectation values of these products on any other state. For example,

$$\langle 0|a_n a_m a_l^\dagger a_s^\dagger|0\rangle = \delta_{ml}\,\delta_{ns} - \delta_{nl}\,\delta_{ms}\,. \qquad (3.37)$$

In the next section, we shall benefit from the use of the Wick theorem.

3.3   Some Thermal Averages

The canonical distribution of a system governed by the Hamiltonian H_0 is given by

$$\rho_0 = \frac{e^{-\beta H_0}}{\mathrm{Tr}\, e^{-\beta H_0}}, \qquad \beta = (k_B T)^{-1}\,. \qquad (3.38)$$

We define the thermal average of the product $a_k^\dagger a_{k'}$ as

$$\langle a_k^\dagger a_{k'}\rangle_0 = \mathrm{Tr}\,\{\rho_0\, a_k^\dagger a_{k'}\}\,. \qquad (3.39)$$

Due to the invariance of the trace under cyclic permutation and using our previous results about the temperature dependence of the operators, see (3.29), the trace in (3.38) is written as

$$\mathrm{Tr}\,\{\rho_0\, a_{k'}\, a_k^\dagger\} = \mathrm{Tr}\,\{a_k^\dagger\,\rho_0\, a_{k'}\} = e^{\beta \mathcal{E}_k}\,\mathrm{Tr}\,\{\rho_0\, a_k^\dagger a_{k'}\}\,, \qquad (3.40)$$

which means that

$$\langle a_{k'}\, a_k^\dagger\rangle_0 = e^{\beta \mathcal{E}_k}\,\langle a_k^\dagger a_{k'}\rangle_0. \qquad (3.41)$$

For fermions, the thermal expectation value of the left-hand side is replaced by

$$\langle a_{k'}\, a_k^\dagger\rangle_0 = \delta_{kk'} - \langle a_k^\dagger a_{k'}\rangle_0\,, \qquad (3.42)$$

from where we get

$$\langle a_k^\dagger a_{k'}\rangle_0 = \frac{\delta_{kk'}}{e^{\beta \mathcal{E}_k} + 1}.$$
(3.43)

An analogous procedure for bosons leads to

$$\langle a_k^\dagger a_{k'}\rangle_0 = \frac{\delta_{kk'}}{e^{\beta \mathcal{E}_k} - 1}.$$
(3.44)

The conservation of the number of particles is enforced by replacing H_0 by $H_0 - \mu N_{op}$, where N_{op} is the number operator $N_{op} = \sum_k a_k^\dagger a_k$. This amounts to the replacement $\mathcal{E}_k \longmapsto \mathcal{E}_k^{(0)} - \mu$ in the expression of the average.

3.4   Interpretation in Terms of Green's Functions

Let us go back to the thermal propagator defined as in (3.5). Consider the Heaviside step function $S(x)$ defined as[1]

$$S(x) := \begin{cases} 1 \text{ if } x > 0, \\ 0 \text{ if } x < 0. \end{cases}$$
(3.45)

The right-hand side of (3.5) can be written as

$$\langle T_\tau[a_k(\tau) a_{k'}^\dagger(\tau')]\rangle = S(\tau - \tau') \langle a_k(\tau) a_{k'}^\dagger(\tau')\rangle - S(\tau' - \tau) \langle a_{k'}^\dagger(\tau') a_k(\tau)\rangle,$$
(3.46)

where the averages are given with respect to some state ρ. Taking into account that

$$a_k(\tau) a_{k'}^\dagger(\tau') = e^{-\mathcal{E}_k \tau} a_k e^{\mathcal{E}_{k'} \tau'} a_{k'}^\dagger = e^{-(\mathcal{E}_k \tau - \mathcal{E}_{k'} \tau')} a_k a_{k'}^\dagger,$$
(3.47)

and taking the averages with respect to ρ_0, we obtain

$$\langle a_k(\tau) a_{k'}^\dagger(\tau')\rangle_0 = e^{-(\mathcal{E}_k \tau - \mathcal{E}_{k'} \tau')} \langle a_k a_{k'}^\dagger\rangle_0 = e^{-(\mathcal{E}_k \tau - \mathcal{E}_{k'} \tau')} (\delta_{kk'} \pm \langle a_{k'}^\dagger a_k\rangle_0)$$
$$= e^{-(\mathcal{E}_k \tau - \mathcal{E}_{k'} \tau')} \delta_{kk'} (1 \pm n_k(\beta)) = e^{-\mathcal{E}_k(\tau-\tau')} \delta_{kk'} (1 \pm n_k(\beta)),$$
(3.48)

where the sign $(+)$ corresponds to bosons and the sign $(-)$ to fermions. The occupation number $n_k(\beta)$ is given by (3.44) for bosons and (3.43) for fermions.

Proceeding analogously, we have that

$$\langle a_{k'}^\dagger(\tau') a_k(\tau)\rangle_0 = e^{\mathcal{E}_k(\tau-\tau')} \delta_{kk'} n_k(\beta).$$
(3.49)

---

[1]Due to the properties of Fourier series and, in particular of their pointwise convergence, one usually chooses $S(0) = 1/2$.

Consequently, Green's function in the case of bosons is given by

$$G_0(k, k', \tau - \tau') = e^{\mathcal{E}_k(\tau - \tau')} \delta_{kk'} \{ S(\tau' - \tau) \, n_k(\beta) - S(\tau - \tau')(1 + n_k(\beta)) \},$$
(3.50)

and for fermions is

$$G_0(k, k', \tau - \tau') = -e^{\mathcal{E}_k(\tau - \tau')} \delta_{kk'} \{ S(\tau' - \tau) \, n_k(\beta) - S(\tau - \tau')(1 - n_k(\beta)) \}.$$
(3.51)

3.4.1 The Truncated Fourier Transform of Green's Function

Take Green's function (3.5) with $k = k'$. We have seen that this choice is the only one for which $G(k, k', \tau - \tau')$ may be different from zero. Since Green's function depends on τ and τ' through their difference, it is also legitimate to use a dependence on the variable $\gamma := \tau - \tau'$. Thus, we shall use the notation $G_0(k, \gamma) \equiv G_0(k, k', \tau - \tau')$. Let $\{\omega_n\}$ be a sequence of real numbers. Then, let us define a truncated Fourier transform for Green's function as

$$\overline{G_0}(k, \omega_n) := \frac{1}{2} \int_{-\beta}^{\beta} d\gamma \, e^{i\omega_n \gamma} \, G_0(k, \gamma).$$
(3.52)

The truncated Fourier transform (3.52) has an inverse that can be written as

$$G_0(k, \gamma) = \frac{1}{\beta} \sum_n e^{-i\omega_n \gamma} \overline{G_0}(k, \omega_n).$$
(3.53)

The proof of (3.53) is quite simple. In fact, if we use (3.53) in the right-hand side of (3.52), we obtain

$$\sum_m \overline{G_0}(k, \omega_m) \left(\frac{1}{2\beta} \int_{-\beta}^{\beta} d\gamma \, e^{i(\omega_n - \omega_m)\gamma} \right) = \sum_m \delta_{nm} \sum_m \overline{G_0}(k, \omega_m) = \overline{G_0}(k, \omega_n). \quad (3.54)$$

This proves our assertion. Then, let us go back to (3.51) valid for fermions and take its truncated Fourier transform (3.52). The result is

$$\overline{G_0}(k, \omega_n) = \frac{1}{2} \int_{-\beta}^{\beta} d\gamma \, e^{i\omega_n \gamma} \, e^{-\mathcal{E}_k \gamma} \{ S(-\gamma) \, n_k(\beta) - S(\gamma)(1 - n_k(\beta)) \}$$

$$= \frac{1}{2} \left[n_k(\beta) \int_{-\beta}^{0} d\gamma \, e^{(i\omega_n - \mathcal{E}_k)\gamma} - (1 - n_k(\beta)) \int_{0}^{\beta} d\gamma \, e^{(i\omega_n - \mathcal{E}_k)\gamma} \right]$$

$$= \frac{1}{2} \left[n_k(\beta) \left(\int_{-\beta}^{0} d\gamma \, e^{-(\mathcal{E}_k - i\omega_n)\gamma} + \int_{0}^{\beta} d\gamma \, e^{-(\mathcal{E}_k - i\omega_n)\gamma} \right) - \int_{0}^{\beta} d\gamma \, e^{-(\mathcal{E}_k - i\omega_n)\gamma} \right]$$

$$= \frac{1}{2} \left[n_k(\beta) \left(\int_{0}^{\beta} d\gamma \, e^{(\mathcal{E}_k - i\omega_n)\gamma} + \int_{0}^{\beta} d\gamma \, e^{-(\mathcal{E}_k - i\omega_n)\gamma} \right) - \int_{0}^{\beta} d\gamma \, e^{-(\mathcal{E}_k - i\omega_n)\gamma} \right].$$
(3.55)

Let us use the notation $x := \mathcal{E}_k - i\omega_n$. After integration over γ, (3.55) becomes

$$\overline{G}_0(k, \omega_n) = \frac{1}{2x} \left(n_k(\beta)[e^{x\beta} - 1 - e^{-x\beta} + 1] + [e^{-x\beta} - 1] \right)$$

$$= \frac{1}{2x} (n_k(\beta)[e^{x\beta} - e^{-x\beta}] + [e^{-x\beta} - 1)]) . \qquad (3.56)$$

So far, we have not made any restriction on the values of ω_n. The idea is to choose ω_n in such a way that (3.56) be as simpler as possible. The choice

$$e^{\pm i\omega_n \beta} = -1 \quad \text{gives} \quad \omega_n = (2n + 1) \frac{\pi}{\beta} . \qquad (3.57)$$

Recall that $e^{x\beta} = e^{\mathcal{E}_k \beta} e^{-i\omega_n \beta}$; thus, a very elementary calculation gives

$$\overline{G}_0(k, (2n + 1) \frac{\pi}{\beta}) = -\frac{1}{x} = \frac{1}{i\omega_n - \mathcal{E}_k} . \qquad (3.58)$$

In order to obtain the same result for bosons, we choose $\omega_n = 2n\pi/\beta$, $n = 0, \pm 1, \pm 2, \dots$.

3.4.2   Summary

- Starting from a given unperturbed Hamiltonian H_0 and defining the normalized density operator

$$\rho_0 \equiv \frac{e^{-\beta H_0}}{\text{Tr}\{e^{-\beta H_0}\}} , \qquad (3.59)$$

the mean value of an observable represented by the operator A, on the state ρ_0, is given by

$$\langle A \rangle \equiv \text{Tr}[\rho_0 A] . \qquad (3.60)$$

- A thermal transformation of magnitude τ is given by

$$A(\tau) \equiv e^{\tau H_0} A e^{-\tau H_0} . \qquad (3.61)$$

In the computation of the above results, a definition of ordered products on τ was introduced by means of the Wick theorem.
- Then, the thermal Green's function for the one particle state reads

$$G_0(k, \tau) := -\langle T_\tau (a_k(\tau_1) a^\dagger(\tau_2)) \rangle_0 , \qquad \tau := \tau_1 - \tau_2 > 0 , \qquad (3.62)$$

and its transform is given by

$$\overline{G}_0(k, \omega_n) = \frac{1}{i\omega_n - \mathcal{E}_k} , \qquad (3.63)$$

with $\mathcal{E}_k = \mathcal{E}_k^0 - \mu$.

- The frequencies

$$\omega_n = (2n + 1)\,\frac{\pi}{\beta}\,, \qquad \text{for fermions}\,,$$

$$\omega_n = 2n\,\frac{\pi}{\beta}\,, \qquad \text{for bosons}\,, \tag{3.64}$$

express the invariance of Green's function under an odd (fermions) or even (fermions) cyclic displacement of the β variable.

3.5 Grand Partition Function

To the unperturbed Hamiltonian H_0, we add the constraint due to the fixed particle number,

$$H_0 \Rightarrow H_0 - \mu N\,, \tag{3.65}$$

which has the shifted eigenvalues $\mathcal{E}_k = \mathcal{E}_k^0 - \mu$, where $\mathcal{E}_k^0$ are the eigenvalues of the original Hamiltonian. To this new constrained Hamiltonian, we add an interaction, λV, where λ is a dimensionless coupling constant. The grand partition function is introduced as usual:

$$\mathcal{Z}(\beta, \mu) := \mathrm{Tr}\,\{e^{-\beta H}\}\,. \tag{3.66}$$

Then, consider the thermal annihilation, $a_k(\tau)$, and creation, $a_k^\dagger(\tau)$, operators defined as in (3.2) and (3.3) and the commutators

$$[H, a_k(\tau)] = [H_0, a_k(\tau)] + \lambda[V, a_k(\tau)]\,. \tag{3.67}$$

So far, we have not made any choice for the potential V. In the simplest case, it represents a two-body interaction. To begin with, let us assume that the particles are fermions. Then, the potential at $\tau = 0$ should be given by an expression of the following form:

$$V = \frac{1}{2} \sum_{k_1, k_2, k_3, k_4} \langle k_1\, k_2 | V | k_3\, k_4 \rangle\, a_{k_1}^\dagger\, a_{k_2}^\dagger\, a_{k_4}\, a_{k_3}\,. \tag{3.68}$$

Next, we study the following commutator:

$$[a_{k_1}^\dagger\, a_{k_2}^\dagger\, a_{k_4}\, a_{k_3}, a_\gamma] = a_{k_1}^\dagger\, a_{k_2}^\dagger\, a_{k_4}\, a_{k_3}\, a_\gamma - a_\gamma\, a_{k_1}^\dagger\, a_{k_2}^\dagger\, a_{k_4}\, a_{k_3}\,. \tag{3.69}$$

Using the anticommuting properties between creation and annihilation operators for fermions in (3.69), we have,

$$[a_{k_1}^\dagger\, a_{k_2}^\dagger\, a_{k_4}\, a_{k_3}, a_\gamma] = a_{k_1}^\dagger\, a_{k_2}^\dagger\, a_{k_4}\, a_{k_3}\, a_\gamma - \delta_{\gamma k_1} a_{k_2}^\dagger\, a_{k_4}\, a_{k_3}$$
$$+ a_{k_1}^\dagger\, \delta_{\gamma k_2}\, a_{k_4}\, a_{k_3} - a_{k_1}^\dagger\, a_{k_2}^\dagger\, a_\gamma\, a_{k_4}\, a_{k_3}\,. \tag{3.70}$$

This gives the commutator between the interaction V and any of the annihilation operators as

$$[V, a_k] = \frac{1}{2} \sum_{k_1, k_2, k_3, k_4} \langle k_1 k_2 | V | k_3 k_4 \rangle \, [a_{k_1}^\dagger a_{k_2}^\dagger a_{k_4} a_{k_3}, a_k]$$

$$= \frac{1}{2} \sum_{k_2, k_3, k_4} \langle k k_2 | V | k_3 k_4 \rangle \, (-a_{k_2}^\dagger a_{k_4} a_{k_3}) + \frac{1}{2} \sum_{k_1, k_3, k_4} \langle k_1 k | V | k_3 k_4 \rangle \, (a_{k_1}^\dagger a_{k_4} a_{k_3}). \quad (3.71)$$

Now, observe that the anticommutation rules give $a_{k_1}^\dagger a_{k_2}^\dagger = -a_{k_2}^\dagger a_{k_1}^\dagger$. Use this fact in (3.68) so as to get

$$V = \frac{1}{2} \sum_{k_1, k_2, k_3, k_4} \langle k_1 k_2 | V | k_3 k_4 \rangle \, a_{k_1}^\dagger a_{k_2}^\dagger a_{k_4} a_{k_3} = -\frac{1}{2} \sum_{k_1, k_2, k_3, k_4} \langle k_1 k_2 | V | k_3 k_4 \rangle \, a_{k_2}^\dagger a_{k_1}^\dagger a_{k_4} a_{k_3}.$$
$$(3.72)$$

Then, take into account that k_1 and k_2 are dummy variables. Consequently, we have that

$$V = \frac{1}{2} \sum_{k_1, k_2, k_3, k_4} \langle k_1 k_2 | V | k_3 k_4 \rangle \, a_{k_1}^\dagger a_{k_2}^\dagger a_{k_4} a_{k_3} = \frac{1}{2} \sum_{k_1, k_2, k_3, k_4} \langle k_2 k_1 | V | k_3 k_4 \rangle \, a_{k_2}^\dagger a_{k_1}^\dagger a_{k_4} a_{k_3}.$$
$$(3.73)$$

Therefore,

$$\langle k_1 k_2 | V | k_3 k_4 \rangle = -\langle k_2 k_1 | V | k_3 k_4 \rangle. \quad (3.74)$$

Next, replace k_1 by k_2 in the last term of (3.71) and use (3.74) in this term. The conclusion is that (3.71) gives

$$[V, a_k] = - \sum_{k_2, k_3, k_4} \langle k k_2 | V | k_3 k_4 \rangle \, a_{k_2}^\dagger a_{k_4} a_{k_3}. \quad (3.75)$$

A similar calculation gives

$$[H_0, a_k] = \sum_{k_1} \mathcal{E}_{k_1} \, [a_{k_1}^\dagger a_{k_1}, a_k] = -\mathcal{E}_k \, a_k. \quad (3.76)$$

Combining (3.75) and (3.76), we finally obtain

$$[H, a_k] = \mathcal{E}_k \, a_k - \lambda \sum_{k_2, k_3, k_4} \langle k k_2 | V | k_3 k_4 \rangle \, a_{k_2}^\dagger a_{k_4} a_{k_3}. \quad (3.77)$$

At this point, we are in the condition to insert the explicit dependence of the operators on τ. Let us define

$$V(\tau) := e^{\tau H} \, V \, e^{-\tau H}. \quad (3.78)$$

Then, we have

$$V(\tau) = \frac{1}{2} \sum_{k_1,k_2,k_3,k_4} \langle k_1 \, k_2 | V | k_3 \, k_4 \rangle \, e^{\tau H} \, a_{k_1}^\dagger \, a_{k_2}^\dagger \, a_{k_4} \, a_{k_3} \, e^{-\tau H}$$

$$= \frac{1}{2} \sum_{k_1,k_2,k_3,k_4} \langle k_1 \, k_2 | V | k_3 \, k_4 \rangle \, (e^{\tau H} \, a_{k_1}^\dagger \, e^{-\tau H})(e^{\tau H} \, a_{k_2}^\dagger \, e^{-\tau H})(e^{\tau H} \, a_{k_4} \, e^{-\tau H})(e^{\tau H} \, a_{k_3} \, e^{-\tau H})$$

$$= \frac{1}{2} \sum_{k_1,k_2,k_3,k_4} \langle k_1 \, k_2 | V | k_3 \, k_4 \rangle \, a_{k_1}^\dagger(\tau) \, a_{k_2}^\dagger(\tau) \, a_{k_4}(\tau) \, a_{k_3}(\tau). \quad (3.79)$$

Next, we take the derivative

$$\frac{\partial}{\partial \tau} \, a_k(\tau) = \frac{\partial}{\partial} \left(e^{\tau H} \, a_k \, e^{-\tau H} \right) = H \, e^{\tau H} \, a_k \, e^{-\tau H} - e^{\tau H} \, a_k \, H \, e^{-\tau H}$$

$$= e^{\tau H} \, [H, a_k] \, e^{-\tau H} = e^{\tau H} \, [H_0 + \lambda V, a_k] \, e^{-\tau H} \,. \quad (3.80)$$

Combining (3.80) with (3.77) leads to

$$\frac{\partial}{\partial \tau} \, a_k(\tau) = -\mathcal{E}_k \, a_k(\tau) - \lambda \sum_{k_2,k_3,k_4} \langle k \, k_2 | V | k_3 \, k_4 \rangle \, a_{k_2}^\dagger(\tau) \, a_{k_4}(\tau) \, a_{k_3}(\tau), \quad (3.81)$$

from where it results the temperature-dependent equation for one particle operator in the presence of interactions:

$$\left(\frac{\partial}{\partial \tau} + \mathcal{E}_k \right) a_k(\tau) = \lambda \sum_{k_2,k_3,k_4} \langle k \, k_2 | V | k_3 \, k_4 \rangle \, a_{k_2}^\dagger(\tau) \, a_{k_4}(\tau) \, a_{k_3}(\tau). \quad (3.82)$$

In (3.82), take $\lambda = 1$, multiply by $a_{k'}^\dagger(\tau')$, sum over k' and take the limits $\tau \longmapsto \tau'$ and $k \longmapsto k'$; then, (3.82) takes the form

$$\frac{1}{2} \lim_{k \to k'} \sum_{k'} \left\{ \lim_{\tau - \tau' \to 0^+} \left(\frac{\partial}{\partial \tau} + \mathcal{E}_k \right) G(k, k', \tau - \tau') \right\} = \langle V \rangle. \quad (3.83)$$

Returning to the expression for the grand partition function, this time using the full Hamiltonian $H(\lambda) = H_0 + \lambda V$, and introducing the Gibbs free energy $\Omega(\lambda)$, one gets

$$\mathcal{Z}(\lambda) = e^{-\beta \Omega(\lambda)} = \mathrm{Tr} \left\{ e^{-\beta H(\lambda)} \right\} = \mathrm{Tr} \left\{ \sum_{l=1}^{\infty} \frac{(-1)^l}{l!} \beta^l \, (H_0 + \lambda V)^l \right\} . \quad (3.84)$$

Taken derivatives with respect to λ

$$\frac{\partial \mathcal{Z}(\lambda)}{\partial \lambda} = \sum_{l=1}^{\infty} \frac{(-1)^l}{l!} \beta^l \frac{\partial}{\partial \lambda} \left\{ \mathrm{Tr} \, (H_0 + \lambda V)^l \right\} , \quad (3.85)$$

or equivalently,

$$\frac{\partial \, \mathcal{Z}(\lambda)}{\partial \, \lambda} = -\frac{\beta}{\lambda} \, e^{-\beta \Omega(\lambda)} \langle \lambda \, V \rangle \,. \tag{3.86}$$

We have multiplied and divided the right-hand side of (3.86) by Tr $\left\{ e^{-\beta H(\lambda)} \right\}$ to get the identity in (3.86). Deriving the expression for $\mathcal{Z}(\lambda)$ in (3.84), with respect to λ, we find that

$$\frac{\partial \, \mathcal{Z}(\lambda)}{\partial \, \lambda} = -e^{-\beta \Omega(\lambda)} \, \beta \, \frac{\partial \, \Omega(\lambda)}{\partial \, \lambda} \,, \tag{3.87}$$

and combining both (3.85) and (3.87), we find that

$$\frac{\partial \, \Omega(\lambda)}{\partial \, \lambda} = \frac{1}{\lambda} \langle \lambda \, V \rangle \,. \tag{3.88}$$

Integrating with respect to λ between 0 and 1, it results that

$$\Omega = \Omega_0 + \int_0^1 d\lambda \, \frac{1}{\lambda} \langle \lambda \, V \rangle \,, \tag{3.89}$$

or

$$\Omega = \Omega_0 + \int_0^1 \frac{d\lambda}{\lambda} \sum_k \lim_{k' \to k} \lim_{\tau - \tau' \to 0^+} \frac{1}{2} \left(\frac{\partial}{\partial \tau} - \mathcal{E}_k \right) G(k, k', \tau - \tau') \,, \tag{3.90}$$

where Ω_0 is the grand potential (Gibbs free energy) corresponding to H_0.

3.5.1   Representation in Terms of Bound States

Let us consider the complete set of bound states, Φ_n, of the operator

$$H \equiv (H_0 - \mu N) + V = \sum_k \mathcal{E}_k \, a_k^\dagger \, a_k + V \,, \tag{3.91}$$

where

$$\mathcal{E}_k = \mathcal{E}_k^0 - \mu N \,. \tag{3.92}$$

The propagator in this basis is written as

$$\begin{aligned}
G(k, k', \tau - \tau') &= -\langle T_\tau \, [a_k(\tau) \, a_{k'}^\dagger(0)] \rangle \\
&= -\sum_n \langle n | e^{-\beta H} \, e^{\tau H} \, a_k \, e^{-\tau H} \, a_{k'}^\dagger | n \rangle \\
&= -\sum_n e^{-\beta E_n} \, e^{\tau E_n} \, \langle n | a_k \, e^{-\tau H} \, a_{k'}^\dagger | n \rangle
\end{aligned}$$

$$= - \sum_{n,n'} e^{-\beta E_n} e^{\tau E_n} \langle n | a_k | n' \rangle \langle n' | e^{-\tau H} a_{k'}^\dagger | n \rangle$$

$$= - \sum_{n,n'} \left(e^{-\beta E_n} e^{\tau (E_n - E_{n'})} \right) \langle n | a_k | n' \rangle \langle n' | a_{k'}^\dagger | n \rangle . \quad (3.93)$$

A discrete Fourier transform of the propagator is then

$$\overline{G}(k, k', \omega_n) := \int_0^\beta d\tau \, e^{i \omega_n \tau} \, G(k, k', \tau)$$

$$= - \sum_{n,n'} \langle n | a_k | n' \rangle \langle n' | a_{k'}^\dagger | n \rangle \int_0^\beta d\tau' \, e^{\tau (E_n - E_{n'}) + i \omega_n \tau} e^{-\beta E_n}$$

$$= - \sum_{n,n'} e^{-\beta E_n} \langle n | a_k | n' \rangle \langle n' | a_{k'}^\dagger | n \rangle \frac{1}{i \omega_n + (E_n - E_{n'})} \left[e^{(i \omega_n + (E_n - E_{n'})) \beta} - 1 \right] . \quad (3.94)$$

For fermions, we have that

$$\omega_n = (2n + 1) \frac{\pi}{\beta} , \quad (3.95)$$

so that (3.94) becomes

$$\overline{G}(k, k', \omega_n) := \sum_{n,n'} \frac{\langle n | a_k | n' \rangle \langle n' | a_{k'}^\dagger | n \rangle}{i \omega_n + (E_n - E_{n'})} \left\{ e^{-\beta E_{n'}} + e^{-\beta E_n} \right\} . \quad (3.96)$$

The physical interpretation of the elements in (3.96) is the following:

(i) The poles of $\overline{G}(k, k', \omega_n)$ represent the excitation energies $E_n - E_{n'}$.
(ii) The weight factors $\langle n | a_k | n' \rangle$ and $\langle n' | a_{k'}^\dagger | n \rangle$ are the amplitudes for the states with one fermion removed or added, respectively.
(iii) The exponential $e^{-\beta E_n}$ is a factor of statistical weights.

3.5.2   Double Green's Functions

In this section, we closely follow the presentation given in the book by Ter Haar [16].
   As we have seen, the time evolution of an observable A is given by the equation

$$A(t) = e^{i H t / \hbar} A \, e^{-i H t / \hbar} , \quad (3.97)$$

where

$$H \longmapsto H - \mu N . \quad (3.98)$$

The canonical average of A is given by

$$\langle A \rangle = \frac{\text{Tr}\,(A\,e^{-\beta H})}{\text{Tr}\,e^{-\beta H}}\,. \tag{3.99}$$

Needless to say that the Hamiltonian H contains interactions.

Definition. Let A and B be two arbitrary operators. The *double Green's function* for these two operators is the correlation function

$$\langle\langle A(t); B(t') \rangle\rangle := \pm i\theta(\mp t \pm t')\left\{ \langle A(t)\,B(t') \rangle - \eta\langle B(t')\,A(t) \rangle \right\}, \tag{3.100}$$

where $\langle - \rangle$ represents the canonical average (3.99), $\eta = \pm$ and $\theta(x)$ is the Heaviside step function, defined as

$$\theta(x) := \begin{cases} 0 \text{ if } x < 0 \\ 1 \text{ if } x > 0 \end{cases}. \tag{3.101}$$

The double Green's function with the upper sign is called *advanced* and with the lower sign is called *retarded*. Thus, the advanced double Green's function for the operators A and B is written as

$$\langle\langle A(t); B(t') \rangle\rangle_a = i\theta(-t + t')\{\langle A(t)\,B(t) \rangle - \eta\,\langle B(t')\,A(t) \rangle\}, \tag{3.102}$$

and the retarded double Green's function is

$$\langle\langle A(t); B(t') \rangle\rangle_r = -i\theta(t - t')\{\langle A(t)\,B(t') \rangle - \eta\,\langle B(t')\,A(t) \rangle\}. \tag{3.103}$$

Both advanced and retarded double Green's functions are Fourier transforms of the kernel $G(\omega)$. More explicitly,

$$\langle\langle A(t), B(t') \rangle\rangle =: \int_{-\infty}^{\infty} d\omega\, G(\omega)\, e^{-i\omega(t-t')}, \tag{3.104}$$

so that

$$G(\omega) := \frac{1}{2\pi} \int_{-\infty}^{\infty} d(t - t')\, e^{i\omega(t-t')}\, \langle\langle A(t); B(t') \rangle\rangle. \tag{3.105}$$

Analogously, we define the functions $J(\omega)$ and $F(\omega)$, respectively, as

$$\langle A(t)\,B(t') \rangle =: \int_{-\infty}^{\infty} d\omega\, J(\omega)\, e^{-i\omega(t-t')}, \tag{3.106}$$

$$\langle B(t')\,A(t) \rangle =: \int_{-\infty}^{\infty} d\omega\, F(\omega)\, e^{-i\omega(t-t')}, \tag{3.107}$$

so that

$$J(\omega) = \frac{1}{2\pi} \int_{-\infty}^{\infty} d(t - t')\, e^{i\omega(t-t')}\, \langle A(t)\,B(t') \rangle, \tag{3.108}$$

and

$$F(\omega) = \frac{1}{2\pi} \int_{-\infty}^{\infty} d(t - t') e^{i\omega(t-t')} \langle B(t') A(t) \rangle. \tag{3.109}$$

The derivative of the double Green's function with respect to t is after (3.104),

$$i \frac{d}{dt} \langle\langle A(t); B(t') \rangle\rangle = \int_{-\infty}^{\infty} d\omega \{\omega G(\omega)\} e^{-i\omega(t-t')}. \tag{3.110}$$

We also have to take the derivative on the right-hand side of (3.100), which is a product of two terms. One is the Heaviside step function, for which its derivative is

$$i \frac{d}{dt} (i\theta(-t + t')) = \delta(t - t'), \tag{3.111}$$

where $\delta(t - t')$ is the Dirac delta function centered at t'. Equation (3.111) is valid for the advanced case. For the retarded case, we have

$$i \frac{d}{dt} (-i\theta(t - t')) = \delta(t - t'), \tag{3.112}$$

hence, the same result. The derivative of the second factor in the right-hand side of (3.100) is

$$\langle i \frac{d}{dt} A(t) B(t') \rangle - \eta \langle B(t') i \frac{d}{dt} A(t) \rangle = \langle [A(t), H] B(t') \rangle - \eta \langle B(t') [A(t), H] \rangle. \tag{3.113}$$

Then, the use of Eqs. (3.100) and (3.110)–(3.113) give

$$i \frac{d}{dt} \langle\langle A(t); B(t') \rangle\rangle = \delta(t - t') \{\langle A(t) B(t') \rangle - \eta \langle B(t') A(t) \rangle\} + \langle\langle [A(t), H]; B(t') \rangle\rangle. \tag{3.114}$$

Relation (3.114) is valid for both the advanced and the retarded functions. We observe that the time evolution of the double Green's function $\langle\langle A(t); B(t') \rangle\rangle$ does not obey a Heisenberg-type equation, since it contains the "source term" including the Dirac delta.

Then, taking into account that the Dirac delta may be written as

$$\delta(t - t') = \frac{1}{2\pi} \int_{-\infty}^{\infty} d\omega\, e^{-i\omega(t-t')}, \tag{3.115}$$

we have that (3.114) becomes

$$\int_{-\infty}^{\infty} d\omega \left[\omega G(\omega) - H(\omega) \right] e^{-i\omega(t-t')}$$
$$= \frac{1}{2\pi} \int_{-\infty}^{\infty} d\omega \langle A(t) B(t') - \eta B(t') A(t) \rangle e^{-i\omega(t-t')}, \tag{3.116}$$

where $G(\omega)$ has been defined in (3.105). Analogously and replacing in (3.105) $\langle\langle A(t); B(t')\rangle\rangle$ by $\langle\langle [A(t), H]; B(t')\rangle\rangle$, we define $H(\omega)$ as the Fourier transform of the latter. In the last row in (3.116), we have taken into account (3.115), the linearity of the averages $\langle -\rangle$ and the independence of these averages on ω. Taking into account that if the Fourier transform of two functions are equal, these functions are also equal, we have that

$$\omega\, G(\omega) - H(\omega) = \frac{1}{2\pi}\,\langle A(t)\,B(t') - \eta\,B(t')\,A(t)\rangle = \frac{1}{2\pi}\,\langle A\,B - \eta\,B\,A\rangle\,.$$
(3.117)

Equation (3.117) shows that $\langle A(t)\,B(t') - \eta\,B(t')\,A(t)\rangle$ cannot depend on t and t'; otherwise, (3.117) must be identically zero. This is why we have written in the right-hand side the expression without this dependence. This relation is valid for either fermions, choosing $\eta = -1$,

$$\omega\, G(\omega) - H(\omega) = \frac{1}{2\pi}\,\langle\{A, B\}\rangle\,,$$
(3.118)

where $\{A, B\} := A\,B + B\,A$ indicates the anticommutator of A and B, or for bosons, choosing $\eta = 1$,

$$\omega\, G(\omega) - H(\omega) = \frac{1}{2\pi}\,\langle [A, B]\rangle\,.$$
(3.119)

These equations determine the structure of $G(\omega)$, which is the Fourier transform of the double Green's function.

Simple Example

Let us assume that the Hamiltonian may be written in a diagonal form in terms of creation and annihilation operators for fermions as

$$H = \sum_l \mathcal{E}_l\, a_l^\dagger\, a_l\,.$$
(3.120)

Then, for two values of l, say k and k', and taken the above operators as $A = a_k$ and $B = a_{k'}^\dagger$, we have that

$$[A, H] = \sum_l \mathcal{E}_l (a_k\, a_l^\dagger\, a_l - a_l^\dagger\, a_l\, a_k)$$

$$= \sum_l \mathcal{E}_l (a_l^\dagger\, a_k\, a_l + a_l\, \delta_{k,l} - a_l^\dagger\, a_k\, a_l) = \sum_l \mathcal{E}_l\, a_l\, \delta_{k,l}\,,$$
(3.121)

and

$$\langle\langle [A, H]; B\rangle\rangle = \mathcal{E}_k\, \langle\langle a_k, a_{k'}^\dagger\rangle\rangle\,.$$
(3.122)

This latter expression has the following consequence:

$$H(\omega) = \frac{1}{2\pi} \int_{-\infty}^{\infty} d(t - t') \, e^{i\omega(t-t')} \, \langle\langle [a_k, H]; a_{k'}^{\dagger} \rangle\rangle$$

$$= \mathcal{E}_k \frac{1}{2\pi} \int_{-\infty}^{\infty} d(t - t') \, e^{i\omega(t-t')} \, \langle\langle a_k; a_{k'}^{\dagger} \rangle\rangle = \mathcal{E}_k \, G(\omega). \qquad (3.123)$$

The last identity in (3.124) comes after the definition of $G(\omega)$ in (3.105) and our choice for A and B. Finally, since the system under consideration is formed by fermions and the grand canonical state is assumed to be normalized, we have that

$$\langle\{a_k, a_{k'}^{\dagger}\}\rangle = \delta_{k,k'}. \qquad (3.124)$$

All these equations finally give for (3.119) the following solution:

$$G(\omega) = \frac{\delta_{k,k'}}{2\pi(\omega - \mathcal{E}_k)}. \qquad (3.125)$$

Next, let us assume that the eigenvalues $|n\rangle$ of the Hamiltonian form a basis for the space of states with $H|n\rangle = \mathcal{E}_n |n\rangle$. The objective is finding a relation between $J(\omega)$ and $F(\omega)$ as defined in (3.108) and (3.109), respectively. Let us first evaluate the following average ($\hbar = 1$):

$$\langle B(t') A(t) \rangle = \text{Tr}\{e^{-\beta H} e^{it'H} B e^{-it'H} e^{itH} A e^{-itH}\}/\text{Tr}\{e^{-\beta H}\}$$

$$= \frac{1}{Z} \sum_{n,m} \langle n|e^{-\beta H} e^{it'H} B|m\rangle\langle m|e^{-it'H} e^{itH} A e^{-itH}|n\rangle$$

$$= \frac{1}{Z} \sum_{n,m} e^{-\beta E_n} e^{it'(E_n - E_m)} \langle n|B|m\rangle e^{it(E_m - E_n)} \langle m|A|n\rangle$$

$$= \frac{1}{Z} \sum_{n,m} e^{-\beta E_n} e^{-i(E_n - E_m)(t - t')} \langle n|B|m\rangle\langle m|A|n\rangle. \qquad (3.126)$$

Then, using (3.109) and (3.126) we have that

$$F(\omega) = \frac{1}{Z} \sum_{n,m} e^{-\beta E_n} \langle n|B|m\rangle\langle m|A|n\rangle \left\{ \frac{1}{2\pi} \int_{-\infty}^{\infty} d(t - t') \, e^{i(\omega - (E_n - E_m))(t - t')} \right\}$$

$$= \frac{1}{Z} \sum_{n,m} e^{-\beta E_n} \langle n|B|m\rangle\langle m|A|n\rangle \, \delta(\omega - E_n + E_m). \qquad (3.127)$$

Then, performing the same operations in (3.108), we have that

$$J(\omega) = \frac{1}{Z} \sum_{n,m} e^{-\beta E_n} \langle n|A|m\rangle\langle m|B|n\rangle \delta(\omega - E_m + E_n). \qquad (3.128)$$

The interchange between the dummy variables n and m does not alter the final result, so that (3.128) must be equal to

$$J(\omega) = \frac{1}{Z} \sum_{n,m} e^{-\beta E_m} \langle n|B|m\rangle \langle m|A|n\rangle \, \delta(\omega - E_n + E_m)$$

$$= e^{\beta\omega} \frac{1}{Z} \sum_{n,m} e^{-\beta E_n} \langle n|B|m\rangle \langle m|A|n\rangle \, \delta(\omega - E_n + E_m) = e^{\beta\omega} F(\omega). \quad (3.129)$$

In summary

$$\boxed{J(\omega) = e^{\beta\omega} F(\omega)}. \quad (3.130)$$

This, along with (3.106) and (3.107), gives

$$\langle A(t)\, B(t') - \eta B(t')\, A(t)\rangle = \int_{-\infty}^{\infty} d\omega \left[(e^{\beta\omega} - \eta)\, F(\omega) \right] e^{-i\omega(t-t')}. \quad (3.131)$$

To obtain the double Green's function, we should multiply equation (3.131) by the factor $-i\theta(t - t')$. Knowing that $\theta(t - t')$ can be written under the integral as the limit,

$$\theta(t - t') = \begin{cases} \lim_{\varepsilon \to 0^+} e^{-\varepsilon(t-t')} & \text{if } t - t' > 0, \\ 0 & \text{if } t - t' < 0. \end{cases} \quad (3.132)$$

Then, using (3.132) in (3.102) and (3.103), we obtain for the *retarded* double Green's function the expression

$$\langle\langle A(t); B(t)\rangle\rangle_r = -i \lim_{\varepsilon \to 0^+} \int_{-\infty}^{\infty} d\omega \left[(e^{\beta\omega} - \eta) F(\omega) \right] e^{-i(\omega - i\varepsilon)(t-t')}. \quad (3.133)$$

The spectral representation of the retarded double Green's function may be defined by multiplying (3.133) by $(2\pi)^{-1} e^{iE(t-t')}$ and integrating from zero to infinity. As a result, we find

$$\langle\langle A; B\rangle\rangle_r^E = -\frac{i}{2\pi} \lim_{\varepsilon \to 0^+} \int_{-\infty}^{\infty} d\omega \int_0^{\infty} d(t - t') \left[(e^{\beta\omega} - \eta) F(\omega) \right] e^{-i(\omega - i\varepsilon)(t-t')} e^{iE(t-t')},$$

$$\quad (3.134)$$

where $\langle\langle A; B\rangle\rangle_r^E$ denotes this *spectral representation for the retarded double Green's function*.

Integration over the variable $u = t - t'$ gives

$$\int_0^{\infty} e^{i(E-\omega+i\varepsilon)(u)} \, du = \frac{1}{i(E - \omega + i\varepsilon)} e^{i(E-\omega)(u)} e^{-\varepsilon(u)} \Big|_0^{\infty}$$

$$= -\frac{1}{i(E - \omega + i\varepsilon)}, \quad (3.135)$$

since $\varepsilon > 0$, so that the second term in (3.135) vanishes at the infinity. As a consequence, the retarded spectral decomposition is

$$\langle\langle A; B \rangle\rangle_r^E = \frac{1}{2\pi} \lim_{\varepsilon \to 0^+} \int_{-\infty}^{\infty} d\omega \, \frac{(e^{\beta\omega} - \eta) \, F(\omega)}{E - \omega + i\varepsilon}. \tag{3.136}$$

The limit in (3.136) cannot be taken inside the integral sign and should be considered as a distribution. This is often written as

$$\langle\langle A; B \rangle\rangle_r^E = \frac{1}{2\pi} \int_{-\infty}^{\infty} d\omega \, \frac{(e^{\beta\omega} - \eta) \, F(\omega)}{E - \omega + i0}. \tag{3.137}$$

Taking in mind the Plemelj formula (3.137) adopts this form

$$\langle\langle A; B \rangle\rangle_r^E = \frac{1}{2\pi} \mathrm{PV} \int_{-\infty}^{\infty} d\omega \, \frac{(e^{\beta\omega} - \eta) \, F(\omega)}{E - \omega} - \frac{i}{2} (e^{\beta E} - \eta) \, F(E)$$
$$= G(E) - \frac{i}{2} (e^{\beta E} - \eta) \, F(E), \tag{3.138}$$

where PV means Cauchy principal value and the last identity defines $G(E)$. Under mild conditions on the function $F(\omega)$, the function in (3.138) admits analytic continuation on the variable E on the upper half plane of the complex plane.

We proceed in a similar manner for the advanced double Green's function, so as to obtain the following result in this case:

$$\langle\langle A; B \rangle\rangle_a^E = \frac{1}{2\pi} \int_{-\infty}^{\infty} d\omega \, \frac{(e^{\beta\omega} - \eta) \, F(\omega)}{E - \omega - i0}$$
$$= G(E) + \frac{i}{2} (e^{\beta E} - \eta) \, F(E). \tag{3.139}$$

The following is a list of the most relevant points, so far introduced.

- Relation between the truncated Fourier transform of double Green's functions:

$$\omega \, G(\omega) - H(\omega) = \frac{1}{2\pi} \langle [A, B]_{\pm} \rangle, \tag{3.140}$$

where the plus sign and the minus sign in (3.140) denote anticommutator (fermions) and commutator (bosons), respectively.
- Spectral functions:

$$\langle A(t) \, B(t') \rangle =: \int_{-\infty}^{\infty} d\omega \, J(\omega) \, e^{-i\omega(t-t')}, \tag{3.141}$$

$$\langle B(t') \, A(t) \rangle =: \int_{-\infty}^{\infty} d\omega \, F(\omega) \, e^{-i\omega(t-t')}, \tag{3.142}$$

which for a discrete spectrum read

$$J(\omega) = \frac{1}{Z} \sum_{n,m} \langle n|A|m \rangle \langle m|B|n \rangle \, e^{-\beta E_n} \, \delta(\omega - E_n + E_m) \,, \qquad (3.143)$$

$$F(\omega) = e^{-\beta \omega} \, J(\omega) \,. \qquad (3.144)$$

- Spectral representations:

$$\langle\langle A; B \rangle\rangle_E = \frac{1}{2\pi} \int_{-\infty}^{\infty} \frac{(e^{\beta \omega} - \eta) F(\omega)}{E - \omega \pm i0}$$

$$= G(E) \mp \frac{i}{2} (e^{\beta E} - \eta) F(E) =: G(E \pm i0) \,,$$

$$\qquad (3.145)$$

where the plus and minus sign in the denominator of the integral in (3.145) stand for retarded and advanced spectral representation, respectively. The function $G(E)$ is given by the integral

$$G(E) = \frac{1}{2\pi} \mathrm{PV} \int_{-\infty}^{\infty} (e^{\beta \omega} - \eta) \, F(\omega) \, \frac{d\omega}{E - \omega} \,, \qquad (3.146)$$

where PV stands for Cauchy principal value.

This notation is fully consistent. Note that, after the definition given in (3.134), the spectral representation for the double Green's function is nothing else that its Fourier transform as given in (3.105), where we have denoted this Fourier transform as $G(\omega)$. This is precisely the function given in (3.145).

3.5.3   Independent Particle Green's Functions

Let us go back to the example of the previous section, in which a Hamiltonian for a system of free fermions is given by (3.120). The eigenvectors of the Hamiltonian form a basis of the space of states. Again, we take $A = a_k$ and $B = a_{k'}^\dagger$ for k and k' being fixed. Then, (3.142) becomes

$$\langle a_{k'}^\dagger(t') a_k(t) \rangle = \int_{-\infty}^{\infty} d\omega \, F(\omega) \, e^{-i\omega(t-t')} \,. \qquad (3.147)$$

This defines $F(\omega)$ for our particular instance. We may find a relation between $F(\omega)$ and $G(\omega \pm i0)$ as defined in (3.145), which is valid in general. From (3.145), one gets (recall that for fermions $\eta = -1$)

$$G(\omega + i0) - G(\omega - i0)$$

$$= \frac{1}{2\pi} \int_{-\infty}^{\infty} (e^{\beta\omega'} + 1) F(\omega') \left[\frac{1}{\omega - \omega' + i0} - \frac{1}{\omega - \omega' - i0} \right] d\omega'$$

$$= G(E) - G(E) - \frac{i}{2} (e^{\beta\omega} + 1) F(\omega) - \frac{i}{2} (e^{\beta\omega} + 1) F(\omega)$$

$$= -i(e^{\beta\omega} + 1) F(\omega), \quad (3.148)$$

where, again, we have made use of (3.145). This gives

$$F(\omega) = \frac{i}{e^{\beta\omega} + 1} \left(G(\omega + i0) - G(\omega - i0) \right). \quad (3.149)$$

Using (3.125), it gives

$$G(\omega + i0) - G(\omega - i0) = \frac{\delta_{k,k'}}{2\pi(\omega - \mathcal{E}_k + i0)} - \frac{\delta_{k,k'}}{2\pi(\omega - \mathcal{E}_k - i0)}. \quad (3.150)$$

Then, from (3.149) and (3.150) in (3.147) we obtain

$$\langle a_{k'}^\dagger(t') a_k(t) \rangle = \frac{\delta_{k,k'}}{2\pi} \int_{-\infty}^{\infty} d\omega \frac{e^{-i\omega(t-t')}}{e^{\beta\omega} + 1} \left[\frac{1}{\omega - \mathcal{E}_k + i0} - \frac{1}{\omega - \mathcal{E}_k - i0} \right]$$

$$= \delta_{k,k'} \frac{e^{-i\mathcal{E}_k(t-t')}}{e^{\mathcal{E}_k \beta} + 1}. \quad (3.151)$$

This formula shows that only terms with $k = k'$ survive. If we write $\mathcal{E}_k = \mathcal{E}_k^0 - \mu$, we arrive to the final expression:

$$\langle a_k^\dagger(t') a_k(t) \rangle = \frac{e^{-i(\mathcal{E}_k^0 - \mu)(t-t')}}{e^{(\mathcal{E}_k^0 - \mu)\beta} + 1}, \quad (3.152)$$

which for $t = t'$ yields the relation between the Green's function and the occupation number, which is,

$$\boxed{G_k(\beta) = [e^{(\mathcal{E}_k^0 - \mu)\beta} + 1]^{-1}}. \quad (3.153)$$

A similar expression for bosons reads

$$\boxed{G_k(\beta) = [e^{(\mathcal{E}_k^0 - \mu)\beta} - 1]^{-1}}. \quad (3.154)$$

3.6   Ferromagnetism

This is another useful example about the use of the above procedure. Let us introduce first some basic ingredients [17].
1. Pauli matrices:

$$S_x = \frac{1}{2} \begin{pmatrix} 0 & 1 \\ 1 & 0 \end{pmatrix}, \qquad S_y = \frac{1}{2} \begin{pmatrix} 0 & -i \\ i & 0 \end{pmatrix}, \qquad S_z = \frac{1}{2} \begin{pmatrix} 1 & 0 \\ 0 & -1 \end{pmatrix}. \tag{3.155}$$

2. Commutation relations between the Pauli matrices:

$$[S_x, S_y] = i\, S_z\,, \qquad [S_z, S_x] = i\, S_y\,, \qquad [S_y, S_z] = i\, S_x\,. \tag{3.156}$$

3. Mapping between creation and annihilation one fermion operator and Pauli matrices:

$$b := S_x + iS_y\,, \quad b^\dagger := S_x - iS_y\,, \quad \{b, b^\dagger\} = 1\,. \tag{3.157}$$

In the case of having more that one particle state, we indicate it by the subindex i. Then, for the ith state, we have the occupation number operator $\mathbf{n}_i := b_i^\dagger b_i$ and the commutation relations

$$[b_i, b_j] = [b_i^\dagger, b_j^\dagger] = 0\,, \qquad [b_i, b_j^\dagger] = \delta_{ij}(1 - 2\mathbf{n}_i)\,. \tag{3.158}$$

4. Pauli matrices in terms of the one particle creation and annihilation operators: by inversion of the mapping (3.157) by means of relations (3.158), we get

$$S_x = \frac{1}{2}(b^\dagger + b)\,, \qquad S_y = \frac{1}{2}(b^\dagger - b)\,, \qquad S_z = -\frac{1}{2}(b^\dagger b - b\, b^\dagger) = \frac{1}{2}(1 - 2\mathbf{n})\,, \tag{3.159}$$

with $\mathbf{n} := b^\dagger b$, where we have assumed the existence of one particle state only.

5. We define the "spin vector" $\mathbf{S}$ as

$$\mathbf{S} := (S_x, S_y, S_z)\,. \tag{3.160}$$

We have one "spin vector" for each particle, which is denoted by $\mathbf{S}_i$. When necessary, we shall write $\mathbf{S}_i = (S_x(i), S_y(i), S_z(i))$.

6. Finally, we have to define a Hamiltonian for a system of spins with an external interaction due to a magnetic field plus an interaction among the particles themselves. This is

$$H = -g\mu_0 B \sum_i S_z(i) - 2 \sum_{ij} I(i, j)\, \mathbf{S}_i \cdot \mathbf{S}_j\,, \tag{3.161}$$

where g, μ_0 and B are constants related with the external magnetic field, the dot in the last term in (3.161) represents a scalar product and $I(i, j)$ are some exchange integrals with the following properties:

$$I(i, i) = 0\,, \qquad I(i, j) = I(j, i)\,, \tag{3.162}$$

for all possible values of i and j. Alternatively, (3.161) can be written in the following equivalent form:

$$H = -g\mu_0 B \sum_i \frac{1}{2}(1 - 2\mathbf{n}_i) - 2\sum_{ij} I(i, j)\, \mathbf{S}_i \cdot \mathbf{S}_j \,, \qquad (3.163)$$

and if N is the total number of particles, we have

$$H = -g\mu_0 B \frac{N}{2} + g\mu_0 B \sum_i \mathbf{n}_i - 2\sum_{ij} I(i, j)\, b_i^\dagger b_j$$

$$-2\sum_{ij} I(i, j)\, \mathbf{n}_i\, \mathbf{n}_j - \frac{N}{2}\sum_j I(j) + 2\sum_i \mathbf{n}_i \sum_j I(j)\,. \qquad (3.164)$$

In the sequel, we shall use the following choices and notation:

(i) Our choice for the operators A and B is $A = b_i$, $B = b_j^\dagger$.

(ii) We may need to use the functions $G(\omega)$ as in (3.105) and the function $H(\omega)$ defined in (3.116) and below for different choices of the operators A and B. One possibility would be to write in general $G(\omega; A, B)$ and $H(\omega; A, B)$. For the choices $A = b_i$, $B = b_j^\dagger$, we simply write $G(\omega; b_i, b_j^\dagger) = G_{ij}(\omega)$ and $H(\omega; b_i, b_j^\dagger) = H_{ij}(\omega)$, for simplicity. For any other choice, we shall use $G(\omega; A, B)$ and $H(\omega; A, B)$.

(iii) n_i denotes the occupation number of the ith level. In the case of fermions, n_i is either 0 or 1. In the case of bosons, this could be any non-negative integer.

If now, we take $\eta = 1$ and use (3.159), we immediately find that

$$F_i G_{ij}(\omega) - \frac{1}{2\pi} \delta_{ij}\, (1 - 2n_i) + H_{ij}(\omega)\,. \qquad (3.165)$$

It remains to give an expression for $G_{ij}(\omega)$ and $H_{ij}(\omega)$. Recall that

$$H_{ij}(\omega) = \frac{1}{2\pi} \int_{-\infty}^{\infty} d(t - t')\, e^{i\omega(t-t')} \langle\langle [b_i, H]; b_j^\dagger \rangle\rangle\,, \qquad (3.166)$$

so that the determination of commutators like $[b_i, H]$ is in order. This gives

$$[b_i, H] = g\mu_0 B \sum_j [b_i, b_j^\dagger b_j] - 2\sum_{kl} I(k, l)\, [b_i, b_k^\dagger b_l]$$

$$-2\sum_{kl} I(k, l)\, [b_i, b_k^\dagger b_k b_l^\dagger b_l] + 2\sum_j I(j) \sum_l [b_i, b_k^\dagger b_k]\,. \qquad (3.167)$$

Commutators in (3.167) are calculated taking into account that $\mathbf{n}_k = b_k^\dagger b_k$ for all values of k:

$$[b_i, b_k^\dagger b_k] = [b_i, b_l^\dagger]\, b_l = \delta_{il}\, (1 - 2\mathbf{n}_i)\, b_i \,, \qquad (3.168)$$

and

$$[b_i, \mathbf{n}_k \, \mathbf{n}_l] = [b_i, \mathbf{n}_k] \, \mathbf{n}_l + \mathbf{n}_k \, [b_i, \mathbf{n}_l] = \delta_{ik} \, (1 - 2\mathbf{n}_i) \, \mathbf{n}_l + \delta_{il} \, \mathbf{n}_k \, (1 - 2\mathbf{n}_i) . \qquad (3.169)$$

Combining the above results, we finally obtain

$$[b_i, H] = g\mu_0 B (1 - 2\mathbf{n}_i) \, b_i - 2 \sum_l I(i, l)(1 - 2\mathbf{n}_i) \, b_l + 2 \sum_j I(j) \, (1 - 2\mathbf{n}_i) \, b_i$$

$$-2 \sum_l I(i, l) \, (1 - 2\mathbf{n}_i) \, \mathbf{n}_l - 2 \sum_k I(k, i) \, \mathbf{n}_k \, (1 - 2\mathbf{n}_i) . \qquad (3.170)$$

When we use (3.170) in (3.166), we have a term of the form

$$H(\omega; \mathbf{n}_l \, b_k, b_j^\dagger) = \frac{1}{2\pi} \int_{-\infty}^{\infty} d(t - t') \, e^{i\omega(t - t')} \, \langle\langle \mathbf{n}_i(t) \, b_k(t); b_j^\dagger(t') \rangle\rangle , \qquad (3.171)$$

the leading contribution to the integral is obtained by applying a mean-field approximation, such that

$$H(\omega; \mathbf{n}_l \, b_k, b_j^\dagger) \approx \langle \mathbf{n}_i \rangle \, H_{ij}(\omega) = n_i \, H_{ij}(\omega) = \bar{n} \, H_{ij}(\omega) . \qquad (3.172)$$

Here, we have written $n_i = \langle \mathbf{n}_i \rangle$. In writing the identity $n_i = \bar{n}$, we have made the assumption that the average occupation numbers is the same for all levels. Then,

$$H_{ij}(\omega) = g\mu_0 B \, G_{ij}(\omega) - 2g\mu_0 B \, \bar{n} \, G_{ij}(\omega)$$

$$-2(1 - 2\bar{n}) \sum_l I(l, i) \, G_{lj}(\omega) + 2(1 - \bar{n}) \sum_{j'} I(j') \, G_{ij}(\omega) . \qquad (3.173)$$

Now, we use (3.173) in (3.166) leading to

$$E \, G_{ij}(\omega) = g\mu_0 B \, G_{ij}(\omega) + 2(1 - \bar{n}) \, I_0 \, G_{ij}(\omega)$$

$$-2(1 - \bar{n}) \sum_l I(l, i) \, G_{lj}(\omega) + \frac{\delta_{ij}}{2\pi} \, 2(1 - \bar{n}) , \qquad (3.174)$$

with

$$I_0 = \sum_{j'} I(j') . \qquad (3.175)$$

Obviously, we may write (3.174) as

$$G_{ij}(\omega)[E - 2(1 - \bar{n}) \, I_0 - g\mu_0 B] = \frac{\delta_{ij}}{2\pi} \, 2(1 - \bar{n}) - 2(1 - \bar{n}) \sum_l I(l, i) \, G_{lj}(\omega) .$$

$$(3.176)$$

Note that (3.176) forms a non-homogeneous linear system for the quantities $G_{lj}(\omega)$ for all values of l. Usually, magnetic spin models use the nearest neighbor approximation, which in our case may be formulated by writing

$$\sum_l I(l, i) \, G_{lj}(\omega) \approx I_q \, G_{ij}(\omega) \, . \tag{3.177}$$

Then, (3.176) and (3.177) together give

$$G_{ij}(\omega)(E - E_q) = \frac{\delta_{ij}}{2\pi} 2(1 - \overline{n}) \, , \tag{3.178}$$

with

$$E_q = 2(1 - \overline{n}) \, I_0 + g\mu_0 B - 2(1 - \overline{n}) \, I_q \, . \tag{3.179}$$

There are (3.179) three terms in the right-hand side of (3.179). The former is the contribution of the spin mean-field, the second one the influence of the external field and the third one the interaction between nearest neighbors. Finally, we arrive at

$$G_{ij}(\omega) = \delta_{ij} \frac{1 - 2\overline{n}}{2\pi(E - E_q)} \, . \tag{3.180}$$

From (3.180), we write

$$\langle b_j^\dagger(t') \, b_k(t) \rangle = \frac{1 - 2\overline{n}}{N} \sum_q \frac{e^{i[(j-k)q - E_q(t-t')]}}{e^{\beta E_q} - 1} \, . \tag{3.181}$$

For $t = t'$, $j = k$ and making the standard approximation of discrete levels by the continuum given by

$$\frac{1}{N} \sum_q \longmapsto \frac{V}{N(2\pi)^3} \int d^3 q \, , \tag{3.182}$$

we have that $\langle b^\dagger(t) \, b(t) \rangle = \overline{n}$, and then

$$\frac{\overline{n}}{1 - 2\overline{n}} = \frac{V}{N} \frac{1}{(2\pi)^3} \int d^3 q \, \frac{1}{e^{\beta E_q} - 1} \, . \tag{3.183}$$

The magnetization σ is obtained from the following relationship:

$$\langle S_z \rangle = \frac{1}{2} (1 - 2\overline{n}) = \frac{\sigma}{2} \, , \tag{3.184}$$

so that (3.183) becomes

$$\frac{\overline{n}}{\sigma} = \frac{V}{N} \frac{1}{(2\pi)^3} \int d^3 q \, \frac{1}{e^{\beta E_q} - 1} \, . \tag{3.185}$$

Path Integrals and Applications

4

In this chapter, we are presenting the basic concepts related to the description of statistical mechanics in terms of amplitudes, rather than probabilities as done in the previous chapters. The proper tool for the purpose of the discussion is the path integral formulation due to Feynman [20–22].

4.1 Basic Ideas

We begin with the time-dependent Schrödinger equation. The wave functions describing a state of one spinless particle in one dimension, under the action of a Hamiltonian H, obey Schroedinger equation:

$$H\psi = i\hbar \frac{\partial \psi}{\partial t}. \tag{4.1}$$

The propagator $G(t, t_0)$ is the solution of the equation

$$\left(H - i\hbar \frac{\partial \psi}{\partial t} \right) G(t, t_0) = -i\hbar I \delta(t - t_0), \tag{4.2}$$

where I is the identity operator. The function $G(t, t_0)$ is called the propagator since it gives the relation between the wave function at time t_0 and at time t by means of

$$\psi(t) = G(t, t_0)\,\psi(t_0). \tag{4.3}$$

If we assume that the Hamiltonian H is independent of time, a solution of (4.2) has the following simple form:

$$G(t, t_0) = \theta(t - t_0)\, e^{-iH(t-t_0)/\hbar}, \tag{4.4}$$

O. Civitarese and M. Gadella, *Methods in Statistical Mechanics*, Lecture Notes
in Physics 974, https://doi.org/10.1007/978-3-030-53658-9_4

where $\theta(t - t_0)$ is the Heaviside step function defined in the previous chapter. If we fix $t_0 = 0$ and $t > 0$, the Heaviside step function is always equal to one. Then, we may define the Green function out of the propagator $G(t, t_0)$ as

$$G(x, y; t) = \langle x | e^{-iHt/\hbar} | y \rangle, \tag{4.5}$$

where the ket $|x\rangle$ is the eigenket of the position operator in one dimension with eigenvalue x. Let us use for some operator A the following formal identity:

$$e^A = \left(e^{A/N} \right)^N, \tag{4.6}$$

where $N = 1, 2, \ldots$ on (4.5), so as to obtain

$$G(x, y; t) = \langle x | \left(e^{-iHt/N\hbar} \right)^N | y \rangle. \tag{4.7}$$

Next, consider the Baker-Hausdorff-Campbell relation: if A and B are a pair of operators, then

$$e^{A+B} = e^A e^B e^{\frac{1}{2}[A,B]} \ldots, \tag{4.8}$$

where $[A, B]$ is the commutator of A and B.

This is a product of exponentials that may be either finite or infinite. Typically if A and B where the position and momentum operators, respectively, their commutator is proportional to the identity operator and therefore the third factor in (4.8) is $e^{i\hbar/2}$, which is a complex number, and the product does not contain any further factor. Then, if one of the factors becomes a number, all the remainder factors become equal to one and the product is finite. Otherwise, it is infinite.

In our case, $H = T + V$, i.e., the Hamiltonian is a sum of two contributions: the kinetic energy T and the potential V. Then, let us write (4.6) as

$$\left\{ \exp\left[-\frac{\lambda}{N}(T + V) \right] \right\}^N = \exp\left[-\frac{\lambda}{N}(T + V) \right] \ldots \exp\left[-\frac{\lambda}{N}(T + V) \right]. \tag{4.9}$$

Define $\lambda := it/\hbar$ and choose $A = \frac{\lambda}{N}T$ and $B = \frac{\lambda}{N}V$. Then, we may apply the Baker-Hausdorff-Campbell formula to each of the factors in the right-hand side of (4.9). In general, T and V do not commute, thus one has to keep all terms in (4.9). However, if we choose large values of N, terms of the order

$$\exp\left[\frac{\lambda^2}{N^2} \frac{1}{2} [T, V] \right] \tag{4.10}$$

and of higher order in $\frac{\lambda}{N}$ may be neglected. Therefore, we approximate (4.9) by

$$\left\{ \exp\left[-\frac{\lambda}{N}(T + V) \right] \right\}^N \approx \exp\left[\frac{-\lambda T}{N} \right] \exp\left[\frac{-\lambda V}{N} \right] \ldots \exp\left[\frac{-\lambda T}{N} \right] \exp\left[\frac{-\lambda V}{N} \right]$$

$$= \left\{ \exp\left[\frac{-\lambda T}{N} \right] \exp\left[\frac{-\lambda V}{N} \right] \right\}^N. \tag{4.11}$$

This approximation becomes exact in the limit $N \longmapsto \infty$ so that

$$G(x, y; t) = \langle x|(e^{-iHt/N\hbar})^N|y\rangle = \lim_{N\to\infty} \langle x|(e^{\lambda T/N} e^{-\lambda V/N})^N|y\rangle. \quad (4.12)$$

The eigenkets of the position operator $|x\rangle$ form a complete set, so that the identity operator may be written as

$$I = \int_{-\infty}^{\infty} dx\, |x\rangle\langle x|. \quad (4.13)$$

Then, for any given operator A, it is always possible to write

$$\langle x|A|y\rangle = \int dx_1 \ldots dx_{N-1} \langle x|A|x_1\rangle\langle x_1|A|x_2\rangle \ldots \langle x_{N-1}|A|y\rangle. \quad (4.14)$$

The integrals

$$I = \int_{-\infty}^{\infty} dx_j\, |x_j\rangle\langle x_j|, \quad (4.15)$$

with $j = 1, 2, \ldots, N - 1$, are copies of (4.13). Using this ideas in (4.12), we obtain the following expression:

$$G(x, y : t) = \lim_{N\to\infty} \int dx_1 \ldots dx_{N-1} \prod_{j=1}^{N-1} \langle x_{j+1}|e^{-\lambda T/N} e^{-\lambda V/N}|x_j\rangle. \quad (4.16)$$

Analogously, the set of eigenkets $|p\rangle$ of the momentum operator also form a complete set, so that the identity operator I may be written as

$$I = \int_{-\infty}^{\infty} dp\, |p\rangle\langle p|. \quad (4.17)$$

As is well known, plane waves with fixed momentum p have the form

$$\varphi_p(\xi) = \frac{1}{\sqrt{2\pi\hbar}} e^{-ip\xi/\hbar} = \langle p|\xi\rangle. \quad (4.18)$$

Then, if $|\eta\rangle$ and $|\xi\rangle$ are two eigenkets of the position operator, we may write

$$\langle \eta|e^{-\lambda T/N}|\xi\rangle = \int_{-\infty}^{\infty} dp\, \langle \eta|e^{-\lambda T/N}|p\rangle\langle p|\xi\rangle$$

$$= \int_{-\infty}^{\infty} dp\, \langle \eta|p\rangle\langle p|\xi\rangle\, e^{-(\lambda/N)p^2/2m} = \frac{1}{2\pi\hbar} \int_{-\infty}^{\infty} dp\, e^{ip(\eta-\xi)/\hbar} e^{-(\lambda/N)p^2/2m} \quad (4.19)$$

The second identity in (4.19) comes from the fact that $T|p\rangle = p^2/(2m)|p\rangle$ and the last one from (4.18).

It is an interesting exercise to solve the integral in the last term in (4.19). This integral has the form

$$ J := \int_{-\infty}^{\infty} e^{-ax^2+bx} \, dx \,, \tag{4.20} $$

with $a > 0$. Then, we write

$$ ax^2 - bx = a(x - x_0)^2 - \frac{b^2}{4a} \,, \quad \text{with} \quad x_0 = \frac{b}{2a} \,, \tag{4.21} $$

to express the exponent as an exact square function of the argument. Then,

$$ J = e^{b^2/(4a)} \int_{-\infty}^{\infty} e^{-a(x-x_0)^2} \, dx = e^{b^2/(4a)} \int_{-\infty}^{\infty} e^{-au^2} \, du = e^{b^2/(4a)} \sqrt{\frac{\pi}{a}} \,, \tag{4.22} $$

where clearly $u := x - x_0$. To solve the last integral in (4.22), let us square it and use polar coordinates:

$$ \left[\int_{-\infty}^{\infty} e^{-au^2} \, du \right]^2 = \left(\int_{-\infty}^{\infty} e^{-au^2} \, du \right) \left(\int_{-\infty}^{\infty} e^{-av^2} \, dv \right) $$

$$ = \int_{-\infty}^{\infty} du \int_{-\infty}^{\infty} dv \, e^{-a(u^2+v^2)} = \int_{0}^{2\pi} d\theta \int_{0}^{\infty} d\rho \, \rho e^{-a\rho^2} = \frac{\pi}{a} \,. \tag{4.23} $$

Going back to (4.19) and taking

$$ a := \frac{\lambda}{2mN} \,, \qquad b = i \frac{\eta - \xi}{\hbar} \,. \tag{4.24} $$

we may perform the integration in (4.19), so that

$$ \langle \eta | e^{-\lambda T/N} | \xi \rangle = \left(\frac{mN}{2\pi \hbar^2 \lambda} \right)^{1/2} \exp \left\{ -\frac{mN}{2\lambda \hbar^2} (\eta - \xi)^2 \right\} \,. \tag{4.25} $$

Let us recall that $|x\rangle$ is an eigenket of the one dimensional position operator $\mathbf{x}$, so that $\mathbf{x}|x\rangle = x|x\rangle$, for any real x. We generally assume that the potential, $V(x)$, is solely a function of the position, and hence, as an operator, it is a function of the position operator of the form $V \equiv V(\mathbf{x})$, so that $V|x\rangle = V(\mathbf{x})|x\rangle = V(x)|x\rangle$. In consequence,

$$ e^{-\lambda V/N} |x_j\rangle = e^{-\lambda V(x_j)/N} |x_j\rangle \,. \tag{4.26} $$

Then, look at the product in the integrand in (4.16) and use (4.26) and (4.25) in the following factor:

$$ \langle x_{j+1} | e^{-\lambda T/N} e^{-\lambda V/N} | x_j \rangle = \langle x_{j+1} | e^{-\lambda T/N} | x_j \rangle e^{-\lambda V(x_j)/N} $$

$$ = \left(\frac{mN}{2\pi \hbar^2 \lambda} \right)^{1/2} \exp \left\{ -\frac{mN}{2\lambda \hbar^2} (x_{j+1} - x_j)^2 - \frac{\lambda}{N} V(x_j) \right\} \,. \tag{4.27} $$

Take $\varepsilon = \hbar \lambda / i N$, then $\lambda / N = i \varepsilon / \hbar$, where ε is real positive number (for construction, λ is purely imaginary since it comes from $\exp[-it(T+V)]$). Also, if λ / N is small, ε will also be small. Then, we may write (4.27) as

$$\langle x_{j+1} | e^{-\lambda T/N} e^{-\lambda V/N} | x_j \rangle = \left(\frac{m}{2 \hbar i \varepsilon} \right)^{1/2} \exp \left\{ \frac{i \varepsilon}{\hbar} \left[\frac{m}{2} \left(\frac{x_{j+1} - x_j}{\varepsilon} \right)^2 - V(x_j) \right] \right\} .$$

(4.28)

This gives the following expression for (4.16):

$$\boxed{G(x, y; t) = \lim_{N \to \infty} \int dx_1 \ldots dx_{N-1} \left(\frac{m}{2 \hbar i \varepsilon} \right)^{N/2} \exp \left\{ \frac{i \varepsilon}{\hbar} \sum_{j=1}^{N-1} \left[\frac{m}{2} \left(\frac{x_{j+1} - x_j}{\varepsilon} \right)^2 - V(x_j) \right] \right\} .}$$

(4.29)

This is an important result as it expresses the propagator in terms of *path integration*. Note that exponential forms of the Hamiltonian, like $\exp[-iHt/\hbar]$, have been replaced by exponential forms of the Lagrangian, like the exponential appearing in (4.29). This is due to the truncations in the original exponential forms, which amounts to the separation between T and V, as in Eq. (4.17).

4.1.1 Interpretation of Results

1. In the limit when the distance between x_j and x_{j+1} for all j goes to zero and $N \longmapsto \infty$, the exponent in the integrand of (4.29) takes the following form:

$$\frac{i}{\hbar} \sum_{j=1}^{N-1} \varepsilon \left[\frac{m}{2} \left(\frac{x_{j+1} - x_j}{\varepsilon} \right)^2 - V(x_j) \right] \approx \frac{i}{\hbar} \int_0^t d\tau \left[\frac{1}{2} m \left(\frac{dx}{d\tau} \right)^2 - V(x) \right]$$

$$= \frac{i}{\hbar} \int_0^t L \, d\tau = \frac{i}{\hbar} S . \quad (4.30)$$

The first term in (4.30) with the sum represents a path starting at the point y at time $t = 0$, which goes through x_1, x_2, etc and ends up with $x = x_N$ at time t. This trajectories may be called discrete. On the other hand, the second term represents a *continuous trajectory* going from $(y, 0)$ to (x, t). The second and third identities define the Lagrangian and the action, respectively. This justifies the idea mentioned before looking at (4.29) as a path integration, as is an integral over all possible discrete paths connecting $(y, 0)$ to (x, t) $t = 0$.

2. With the notation

$$\exp \left\{ \frac{i}{\hbar} S([x]) \right\} , \quad (4.31)$$

we mean that the action S should be calculated along one unique discrete trajectory. By

$$\sum_{[x]} \exp\left\{\frac{i}{\hbar} S([x])\right\}, \tag{4.32}$$

we mean "summation" over all possible discrete trajectories. Then, (4.32) defines the *path integral*.

3. Observe that the product

$$\left[\left(\frac{m}{2\hbar i\varepsilon}\right)^{1/2} \cdot \left(\frac{m}{2\hbar i\varepsilon}\right)^{1/2} \cdot \cdots \cdot \left(\frac{m}{2\hbar i\varepsilon}\right)^{1/2}\right] \left\{\begin{array}{c} N \text{ terms}, \ N \to \infty \\ \varepsilon \to 0 \end{array}\right\} \tag{4.33}$$

that appears under the integral sign in (4.29) diverges in the limits $N \longmapsto \infty$, $\varepsilon \longmapsto 0$. This product has the meaning of an element of the metric, so that the path integration will be defined and finite for any quadratic function of the coordinates and momenta.

4. The propagator is precisely the path integral (4.32),

$$G(x, y; t) = \sum_{[x]} \exp\left\{\frac{i}{\hbar} S([x])\right\}. \tag{4.34}$$

The stationary phase method gives an asymptotic formula for some oscillatory integrals of the type

$$\int_{-\infty}^{\infty} dt \, g(t) \, e^{i\lambda f(t)}, \tag{4.35}$$

for large values of the real constant λ and for some conditions on the function $g(t)$. In particular, if $g(t) \equiv 1$ for large λ we have the following approximation:

$$F(\lambda) = \int_{-\infty}^{\infty} dt \, e^{i\lambda f(t)} \approx \sum_{k} e^{i\lambda f(t_k)} \sqrt{\frac{2\pi}{\lambda |f''(t_k)|}}, \tag{4.36}$$

where the sum extends to all points t_k such that $f'(t_k) = 0$ and $f''(t_k) \neq 0$.

4.2   Path Integrals with Vector Potentials

Let us consider a path integration over continuous paths. We reach the points y and x at times $t = 0$ and t, respectively. We denote this path integral as

$$G(x, y; t) = \int_{(y,0)}^{(x,t)} d[x(\tau)] \, e^{iS[x(\tau)]/\hbar}, \tag{4.37}$$

where

$$S[x(\tau)] = \int_0^t L\left(x, \frac{dx}{d\tau}\right) d\tau, \tag{4.38}$$

and L is the Lagrangian:

$$L = \frac{1}{2}m \left(\frac{dx}{d\tau}\right)^2 - V(x). \tag{4.39}$$

Adding to the Lagrangian a vector potential, $\mathbf{A}(\mathbf{x})$, we write

$$L = \frac{1}{2}m \left(\frac{d\mathbf{x}}{d\tau}\right)^2 + \frac{e}{c}\left(\frac{d\mathbf{x}}{d\tau}\right) \cdot \mathbf{A}(\mathbf{x}) - V(\mathbf{x}), \tag{4.40}$$

where the dot means scalar product, e is the particle charge and c is the speed of light in the vacuum, respectively. The contribution to the action due to the vector field is then written as

$$\frac{e}{c} \int_0^t \left(\frac{d\mathbf{x}}{d\tau}\right) \cdot \mathbf{A}(\mathbf{x}) \, d\tau = \frac{e}{c} \int_{\mathbb{R}^3} A(\mathbf{x}) \cdot d\mathbf{x}. \tag{4.41}$$

The discrete equivalent of (4.41) is the discrete path $y, x_1, x_2, \ldots, x_N, x$ with N intermediate steps

$$\frac{e}{c} \sum_{j=1}^{N} (\mathbf{x}_{j+1} - \mathbf{x}_j) \cdot \mathbf{A}\left(\frac{\mathbf{x}_{j+1} + \mathbf{x}_j}{2}\right). \tag{4.42}$$

In (4.42), the potential vector has to be taken at some point related to $\mathbf{x}_{j+1}$ and $\mathbf{x}_j$ and we have chosen the mean point between both. Then in (4.29), we have to add this term into the exponent under the integral sign. We obtain

$$G(x, y; t) = \lim_{N \to \infty} \int d^3\mathbf{x}_1 \ldots d\mathbf{x}_N \left(\frac{m}{2\pi i \hbar \varepsilon}\right)^{3(N+1)/2}$$

$$\times \exp\left\{\frac{i\varepsilon}{\hbar} \sum_{j=0}^{N} \left[\frac{m}{2}\left(\frac{\mathbf{x}_{j+1} - \mathbf{x}_j}{\varepsilon}\right)^2 - V(\mathbf{x}_j) + \frac{e}{c} \sum_{j=1}^{N}(\mathbf{x}_{j+1} - \mathbf{x}_j) \cdot \mathbf{A}\left(\frac{\mathbf{x}_{j+1} + \mathbf{x}_j}{2}\right)\right]\right\}. \tag{4.43}$$

As an example, let us take the wave function at time T, $\psi(\mathbf{y}, T)$ and use the propagator to calculate the wave function at time $T + \varepsilon$ at the point $\mathbf{x}$, $\psi(\mathbf{x}, T + \varepsilon)$. To this end, we write

$$\psi(\mathbf{x}, T + \varepsilon) = \int_{\mathbb{R}^3} d^3\mathbf{y} \left(\frac{m}{2\pi i \hbar \varepsilon}\right)^{3/2} \exp\left\{\frac{i\varepsilon}{\hbar}\left[\frac{m}{2}\left(\frac{\mathbf{x} - \mathbf{y}}{\varepsilon}\right)^2 - V(\mathbf{y})\right]\right.$$

$$\left. + \frac{ie}{\hbar c}(\mathbf{x} - \mathbf{y}) \cdot \mathbf{A}\left(\frac{\mathbf{x} + \mathbf{y}}{2}\right)\right\} \psi(\mathbf{y}, T). \tag{4.44}$$

Expanding the exponents in (4.44) and considering small displacements $\boldsymbol{\xi} := \mathbf{y} - \mathbf{x}$, we obtain

$$V(\mathbf{y}) = V(\mathbf{x} + \boldsymbol{\xi}) = V(\mathbf{x}) + \frac{\partial V(\mathbf{x})}{\partial x_i} \xi_i + \cdots = V(\mathbf{x}) + \boldsymbol{\nabla} V(\mathbf{x}) \cdot \boldsymbol{\xi} + \dots,$$

(4.45)

$$(\mathbf{x} - \mathbf{y}) \cdot \mathbf{A}\left(\frac{\mathbf{x} + \mathbf{y}}{2}\right) = -\boldsymbol{\xi} \cdot \mathbf{A}(\mathbf{x}) - \frac{1}{2} (\boldsymbol{\xi} \cdot \boldsymbol{\nabla})\mathbf{A}(\mathbf{x}) + \dots.$$

(4.46)

Inserting these expressions in (4.44),

$$\frac{i\varepsilon}{\hbar}\left[\frac{m}{2}\left(\frac{\mathbf{x} - \mathbf{y}}{\varepsilon}\right)^2 - V(\mathbf{y})\right] + \frac{ie}{\hbar c}(\mathbf{x} - \mathbf{y}) \cdot \mathbf{A}\left(\frac{\mathbf{x} + \mathbf{y}}{2}\right)$$

$$= \frac{i\varepsilon}{\hbar}\frac{m}{2}\frac{\boldsymbol{\xi}^2}{\varepsilon^2} - \frac{i\varepsilon}{\hbar}[V(\mathbf{x}) + \boldsymbol{\nabla} V(\mathbf{x}) \cdot \boldsymbol{\xi} + \dots] - \frac{ie}{\hbar c}\left[\boldsymbol{\xi} \cdot \mathbf{A}(\mathbf{x}) + \frac{1}{2}(\boldsymbol{\xi} \cdot \boldsymbol{\nabla})\mathbf{A}(\mathbf{x}) + \dots\right].$$

, (4.47)

and expanding the wave function as a Taylor series, we write

$$\psi(\mathbf{y}, T) = \psi(\mathbf{x} + \boldsymbol{\xi}, T) = \psi(\mathbf{x}, T) + \boldsymbol{\xi} \cdot \boldsymbol{\nabla} \psi(\mathbf{x}, T) + \frac{1}{2}\sum_{n,m}\xi_n\xi_m\frac{\partial^2 \psi}{\partial x_m \partial x_m} + \dots.$$

(4.48)

Next, we take all the above expansions and replace them in (4.44), retaining lower order terms, to write

$$\psi(\mathbf{x}, T + \varepsilon) = \left(\frac{m}{2\pi i \hbar\varepsilon}\right)^{3/2} \int_{\mathbb{R}^3} d^3\boldsymbol{\xi} \, \exp\left\{\frac{im\boldsymbol{\xi}^2}{2\varepsilon \hbar}\right\}$$

$$\times \left(1 - \frac{i\varepsilon}{\hbar} V(\mathbf{x}) + \frac{1}{2}\left(-\frac{i\varepsilon}{\hbar} V(\mathbf{x})\right)^2 + \dots\right.$$

$$\left. -\frac{i\varepsilon}{\hbar}\boldsymbol{\nabla} V(\mathbf{x}) \cdot \boldsymbol{\xi} + \frac{1}{2}\left(\frac{i\varepsilon}{\hbar}\boldsymbol{\nabla} V(\mathbf{x}) \cdot \boldsymbol{\xi}\right)^2 + \dots\right)$$

$$\times \left(1 - \frac{ie}{\hbar c}\boldsymbol{\xi} \cdot \mathbf{A}(\mathbf{x}) + \frac{1}{2}\left(\frac{ie}{\hbar c}\boldsymbol{\xi} \cdot \mathbf{A}(\mathbf{x})\right)^2 + \dots\right.$$

$$\left. -\frac{ie}{\hbar c}\frac{1}{2}(\boldsymbol{\xi} \cdot \boldsymbol{\nabla})\mathbf{A}(\mathbf{x}) + \frac{1}{2}\left(\frac{ie}{\hbar c}\frac{1}{2}(\boldsymbol{\xi} \cdot \boldsymbol{\nabla})\mathbf{A}(\mathbf{x})\right)^2 + \dots\right)$$

$$\times (\psi(\mathbf{x}, T) + \boldsymbol{\xi} \cdot \boldsymbol{\nabla} \psi(\mathbf{x}, T) + \dots).$$

(4.49)

Then, keeping leading order terms in $\boldsymbol{\xi}$ and reordering, we write

$$
\psi(\mathbf{x}, T + \varepsilon) = \left(\frac{m}{2\pi i\,\hbar\varepsilon}\right)^{3/2} \int_{\mathbb{R}^3} d^3\boldsymbol{\xi}\, e^{im\xi^2/(2\hbar\varepsilon)} \left(1 - \frac{i\varepsilon}{\hbar} V(\mathbf{x})\right)
$$

$$
\left(1 - \frac{ie}{\hbar c}\,\xi_n A_n - \frac{ie}{2\hbar c}\,\xi_n \xi_m \partial_m A_n - \frac{1}{2}\frac{e^2}{\hbar^2 c^2}\,\xi_n A_n \xi_m A_m\right)
$$

$$
\times \left(\psi + \xi_l\,\partial_l\,\psi + \frac{1}{2}\,\xi_l \xi_k\,\partial_l\,\partial_k\,\psi\right), \quad (4.50)
$$

where for the summations over repeated indices we have adopted the notation: $\sum_n a_n b_n \equiv a_n b_n$, $\partial_k := \partial/(\partial x_k)$, $\mathbf{A} = (A_1, A_2, A_3)$, etc. Reordering the right-hand side of the equation in terms of the wave function and its derivatives, one gets

$$
\left(\frac{m}{2\pi i\,\hbar\varepsilon}\right)^{3/2} \int_{\mathbb{R}^3} d^3\boldsymbol{\xi}\, e^{im\xi^2/(2\hbar\varepsilon)} \left(1 - \frac{i\varepsilon}{\hbar} V(\mathbf{x})\right)
$$

$$
\left(1 - \frac{ie}{\hbar c}\,\xi_n A_n - \frac{ie}{2\hbar c}\,\xi_n \xi_m \partial_m A_n - \frac{1}{2}\frac{e^2}{\hbar^2 c^2}\,\xi_n A_n \xi_m A_m\right) \psi
$$

$$
\left(1 - \frac{ie}{\hbar c}\,\xi_n A_n - \frac{ie}{2\hbar c}\,\xi_n \xi_m \partial_m A_n - \frac{1}{2}\frac{e^2}{\hbar^2 c^2}\,\xi_n A_n \xi_m A_m\right) \xi_l\,\partial_l\,\psi
$$

$$
\left(1 - \frac{ie}{\hbar c}\,\xi_n A_n - \frac{ie}{2\hbar c}\,\xi_n \xi_m \partial_m A_n - \frac{1}{2}\frac{e^2}{\hbar^2 c^2}\,\xi_n A_n \xi_m A_m\right) \frac{1}{2}\,\xi_l \xi_k\,\partial_l\,\partial_k\,\psi. \quad (4.51)
$$

Equation (4.51) is a sum of products of integrals of the form

$$
\int_{-\infty}^{\infty} e^{-ax^2}\, x^r\, dx, \quad (4.52)
$$

which vanishes for odd values of r. Therefore, (4.51) becomes

$$
\psi(\mathbf{x}, T + \varepsilon) = \left(\frac{m}{2\pi i\,\hbar\varepsilon}\right)^{3/2} \int_{\mathbb{R}^3} d^3\boldsymbol{\xi}\, e^{im\xi^2/(2\hbar\varepsilon)} \left(1 - \frac{i\varepsilon}{\hbar} V(\mathbf{x})\right)
$$

$$
\times \left(\psi - \frac{ie}{2\hbar c}\,\xi_n \xi_m \partial_m A_n\,\psi - \frac{1}{2}\frac{e^2}{\hbar^2 c^2}\,\xi_n A_n \xi_m A_m\,\psi - \frac{ie}{\hbar c}\,\xi_n A_n \xi_l\,\partial_l\,\psi + \frac{1}{2}\,\xi_l \xi_k\,\partial_l\,\partial_k\,\psi\right).
$$

$$
(4.53)
$$

The non-vanishing integrals are ($a = -im/(2\varepsilon\,\hbar)$)

$$
\int_{\mathbb{R}^3} d^3\boldsymbol{\xi}\, e^{im\xi^2/(2\varepsilon\,\hbar)} = \left(\int_{-\infty}^{\infty} d\xi\, e^{-a\xi^2}\right)^3 = \left(\sqrt{\frac{2i\pi\varepsilon\hbar}{m}}\right)^3, \quad (4.54)
$$

and

$$
\int_{\mathbb{R}^3} d^3\boldsymbol{\xi}\, \xi_k \xi_l\, e^{im\xi^2/(2\varepsilon\,\hbar)} = \delta_{kl} \left(\sqrt{\frac{2i\pi\varepsilon\hbar}{m}}\right)^2 \sum_{i=1}^3 \int_{-\infty}^{\infty} d\xi_i\, \xi_i^2\, e^{im\xi^2/(2\varepsilon\,\hbar)}
$$

$$
= \delta_{kl}\,\frac{i\varepsilon\hbar}{m} \left(\sqrt{\frac{2i\pi\varepsilon\hbar}{m}}\right)^3. \quad (4.55)
$$

In consequence,

$$\psi(\mathbf{x}, T + \varepsilon) = \left(1 - \frac{i\varepsilon}{\hbar} V(\mathbf{x})\right) \psi(\mathbf{x}, T)$$

$$+\varepsilon \left[\frac{e}{2mc} (\nabla \cdot \mathbf{A}) - \frac{ie^2}{\hbar mc^2} \mathbf{A}^2 \frac{e}{mc} (\mathbf{A} \cdot \nabla) + \frac{i\hbar}{2m} \nabla^2\right] \psi(\mathbf{x}, T), \qquad (4.56)$$

from where, it results

$$[\psi(\mathbf{x}, T + \varepsilon) - \psi(\mathbf{x}, T)] =$$

$$= \frac{\varepsilon}{\hbar} \left\{ i \frac{\hbar^2}{2m} \nabla^2 - i V(\mathbf{x}) - i \frac{e^2}{mc^2} \mathbf{A}^2 - \frac{e\hbar}{mc} (\mathbf{A} \cdot \nabla + \nabla \cdot \mathbf{A}) \right\} \psi(\mathbf{x}, T).$$
$$(4.57)$$

In consequence,

$$i\hbar \left[\frac{\psi(\mathbf{x}, T + \varepsilon) - \psi(\mathbf{x}, T)}{\varepsilon}\right]$$

$$= -\frac{\hbar^2}{2m} \nabla^2 \psi + V(\mathbf{x}) \psi + \frac{e^2}{mc^2} \mathbf{A}^2 \psi + \frac{ie\hbar}{mc} (\mathbf{A} \cdot \nabla) \psi + \frac{ie\hbar}{mc} \psi \nabla \cdot \mathbf{A}.$$
$$. (4.58)$$

Taking the limit $\varepsilon \longmapsto 0$, we finally obtain

$$i\hbar \frac{\partial \psi}{\partial t} = \frac{1}{2m} \left(-i\hbar \nabla - \frac{e}{c} \mathbf{A}\right)^2 \psi + V\psi, \qquad (4.59)$$

which is the Schrödinger equation of a charged particle moving in a vector potential $\mathbf{A}(\mathbf{x})$.

4.3   A Working Example

The propagator of a free particle in one dimension is

$$G(x, y; t) = \int_{(y,0)}^{(x,t)} dx(\tau) \exp\left\{\frac{i}{\hbar} \int_0^t \frac{m}{2} \left(\frac{dx}{d\tau}\right)^2 d\tau\right\}$$

$$= \lim_{N \to \infty} \lim_{\varepsilon \to 0} \left(\frac{m}{2\pi i \hbar \varepsilon}\right)^{(N+1)/2} \int dx_1 \dots dx_N \exp\left\{\frac{im}{2\hbar\varepsilon} \sum_{j=1}^{N} (x_{j+1} - x_j)^2\right\}, \quad (4.60)$$

where the path has $N + 1$ steps, i.e., $y = x_0, x_1, \dots, x_N, x_{N+1} = x$ and $t = \varepsilon(N + 1)$. The integrals in the second row of (4.60) have the form,

$$\int_{-\infty}^{\infty} du \, e^{-a(x-u)^2 - b(u-y)^2}. \qquad (4.61)$$

Simple manipulations give

$$a(x - u)^2 + b(u - y)^2 = ax^2 + by^2 + (a + b)u^2 - (2ax + 2by)u . \quad (4.62)$$

Using (4.62) in (4.61), we obtain

$$e^{-ax^2 - by^2} \int_{-\infty}^{\infty} du \, e^{-(a+b)u^2 + 2(ax+by)u} . \quad (4.63)$$

This type of integrals have been solved before (see (4.23)). Using the same procedure, we obtain for (4.61) and, hence, for (4.63) the following expression:

$$\int_{-\infty}^{\infty} du \, e^{-a(x-u)^2 - b(u-y)^2} = e^{-ax^2 - by^2} \sqrt{\frac{\pi}{a+b}} \, e^{(ax+by)^2/(a+b)}$$

$$= \sqrt{\frac{\pi}{a+b}} \, e^{\frac{-(a+b)(ax^2+by^2)+a^2x^2+b^2y^2+2(abxy)}{a+b}} = \sqrt{\frac{\pi}{a+b}} \, e^{-\frac{ab}{a+b}(x-y)^2} . \quad (4.64)$$

Then, we are in the position of solving the integrals in the second row of (4.60). For a and b, we choose

$$a = b = -\frac{im}{2\hbar\varepsilon} \implies \frac{ab}{a+b} = -\frac{im}{4\hbar\varepsilon} , \quad (4.65)$$

so that, for the integration on x_N, we have

$$\left(\frac{m}{2\pi i \, \hbar\varepsilon}\right) \int_{-\infty}^{\infty} dx_N \exp\left\{\frac{im}{2\hbar\varepsilon}[(x - x_N)^2 + (x_N - x_{N-1})^2]\right\}$$

$$= \left(\frac{m}{2\pi i \, \hbar\varepsilon}\right) \sqrt{\frac{i\pi \, \hbar\varepsilon}{m}} \exp\left\{\frac{im}{2\hbar(2\varepsilon)}(x - x_{N-1})^2\right\}$$

$$= \sqrt{\frac{m}{2\pi i \, \hbar(2\varepsilon)}} \exp\left\{\frac{im}{2\hbar(2\varepsilon)}(x - x_{N-1})^2\right\} . \quad (4.66)$$

Then, we perform the integration over the variable x_{N-1} applying the same type of procedure. Nevertheless, we have to take into account the result obtained after the first integral. Then, we calculate the product of (4.66) with the integral on x_{N-1}, noticing that now

$$a = -\frac{im}{2\hbar(2\varepsilon)} , \qquad b = -\frac{im}{2\hbar\varepsilon} , \quad (4.67)$$

which gives

$$\sqrt{\frac{m}{2\pi i \, \hbar(2\varepsilon)}} \exp\left\{\frac{im}{2\hbar(2\varepsilon)}(x - x_{N-1})^2\right\} \sqrt{\frac{m}{2\pi i \, \hbar(2\varepsilon)}} \exp\left\{\frac{im}{2\hbar(\varepsilon)}(x - x_{N-1})^2\right\}$$

$$\times \int_{-\infty}^{\infty} dx_{N-1} \exp\left\{\frac{im}{2\hbar(2\varepsilon)}(x - x_{N-1})^2 + \frac{im}{2\hbar\varepsilon}(x_{N-1} - x_{N-2})^2\right\}$$

$$= \sqrt{\frac{m}{2\pi i \, \hbar(3\varepsilon)}} \exp\left\{\frac{im}{2\hbar(3\varepsilon)}(x - x_{N-2})^2\right\} . \quad (4.68)$$

Then, we multiply (4.68) times the result of performing the integral over the variable x_{N-2}. The result is

$$\sqrt{\frac{m}{2\pi i\,\hbar(4\varepsilon)}}\,\exp\left\{\frac{im}{2\hbar(4\varepsilon)}\,(x-x_{N-2})^2\right\}. \tag{4.69}$$

Note that the number multiplying ε increases in one unit at each step. After the last integral, we have made N steps, so that this factor will be just $\varepsilon(N+1)$. Taking into account that $t=\varepsilon(N+1)$, the final result is

$$\boxed{G(x,y;t)=\sqrt{\frac{m}{2\pi i\,\hbar t}}\,\exp\left\{\frac{im}{2\hbar t}\,(x-y)^2\right\},} \tag{4.70}$$

which is the propagator for the free particle. The classical motion of a one-dimensional free particle from an initial point y to a final point x is

$$x(\tau)=y+\frac{\tau}{t}\,(x-y), \tag{4.71}$$

and, consequently, the action of this *classical* particle is given by

$$S_{\text{cl}}=\int_0^t L\,d\tau=\int_0^t \frac{m}{2}\left(\frac{dx(\tau)}{d\tau}\right)^2 d\tau=\int_0^t \frac{m}{2}\left(\frac{x-y}{t}\right)^2 d\tau=\frac{m}{2}\frac{(x-y)^2}{t}. \tag{4.72}$$

Now, comparing (4.72) to (4.70), we have that

$$G(x,y;t)=\sqrt{\frac{m}{2\pi i\,\hbar t}}\,\exp\left\{\frac{i}{\hbar}\,S_{\text{cl}}\right\}. \tag{4.73}$$

4.4 Applications to Statistical Mechanics

4.4.1 Particle Interaction with a Central Potential

Let us consider a system of N identical particles interacting with a common potential $V(x)$, although not interacting among themselves. The total partition function is the product of the partition function of each particle, so that

$$Z=\left[\text{Tr}\left\{e^{-\beta H}\right\}\right]^N=\left[\text{Tr}\left\{e^{-i(\beta\hbar/i)\,H/\hbar}\right\}\right]^N=\left[\int dx\,G(x,x;-i\beta\hbar)\right]^N, \tag{4.74}$$

where $-i\beta\hbar$ plays the role of t. The one particle Hamiltonian is $H=p^2/(2m)+V(x)$, so that the propagator is given by ($t=-i\beta\hbar$):

$$G(x,y;t)=\left(\frac{m}{2\pi i\,\hbar t}\right)^{1/2}\exp\left\{\frac{i}{\hbar}\left(\frac{m}{2}\frac{(x-y)^2}{t}-V(x)t\right)\right\}. \tag{4.75}$$

Applying this result in (4.74) and taking into account that in (4.74) we should integrate in $x = y$, the resulting partition function looks like

$$Z^{1/N} = \frac{1}{\hbar} \left(\frac{m}{2\pi\beta} \int dx \, e^{-\beta V(x)} \right),$$ (4.76)

which is the first contribution to the partition function due to the common potential. The integral in (4.75) is performed by neglecting the volume of the particles, v_0. Otherwise, to include this effect, we may exclude from the integral the proper volume Nv_0.

4.4.2 Representation of the Partition Function Using Coherent States: The Harmonic Oscillator in the Number Representation

Let a and $a^\dagger$ be the annihilation and creation operators, which act on a vacuum $|0\rangle$, such that $a \, |0\rangle = 0$ and

$$|n\rangle = \frac{(a^\dagger)^n}{\sqrt{n!}} \, |0\rangle, \qquad n = 0, 1, 2, \ldots$$ (4.77)

are the eigenstates of the harmonic oscillators.

Then, for any complex number z, we define a coherent state with normalization constant $N(z) = e^{-|z|^2/2}$ as

$$|z\rangle := N(z) \, e^{z a^\dagger} \, |0\rangle, \qquad \text{with}.$$ (4.78)

The coherent state $|z\rangle$ is an eigenstate of the annihilation operator a with eigenvalue z, i.e.,

$$a|z\rangle = N(z) \sum_{n=0}^{\infty} \frac{z^n}{n!} \, a \, (a^\dagger)^n \, |0\rangle,$$ (4.79)

and since $[a, (a^\dagger)^n] = n(a^\dagger)^{n-1}$, we readily obtain that $a|z\rangle = z \, |z\rangle$.

Coherent states fulfill the following completeness condition, if $z = u + iv$:

$$I = \int_{\mathbb{R}^2} du \, dv \, \frac{1}{\pi} \, |z\rangle\langle z| = \int_{\mathbb{C}} d^2z \, \frac{1}{\pi} \, |z\rangle\langle z|,$$ (4.80)

where I is the identity operator. These are two different notations for the same relation. Another is

$$\int_{\mathbb{C}} dz \, (z^*)^n \, z^m \, e^{-|z|^2} = \pi \, n! \, \delta_{nm},$$ (4.81)

where the asterisk denotes complex conjugation.

The trace of a one particle operator A in the one dimension harmonic oscillator basis is

$$\text{Tr } A = \sum_{n=0}^{\infty} \langle n|A|n \rangle \tag{4.82}$$

and inserting the identity (4.80) twice, we have that

$$\sum_{n=0}^{\infty} \int_{\mathbb{C}} \frac{d^2\alpha}{\pi} \int_{\mathbb{C}} \frac{d^2\beta}{\pi} \langle n|\alpha \rangle \langle \alpha|A|\beta \rangle \langle \beta|n \rangle = \sum_{n=0}^{\infty} \int_{\mathbb{C}} \frac{d^2\alpha}{\pi} \int_{\mathbb{C}} \frac{d^2\beta}{\pi} \langle \beta|n \rangle \langle n|\alpha \rangle \langle \alpha|A|\beta \rangle$$

$$= \int_{\mathbb{C}} \frac{d^2\alpha}{\pi} \int_{\mathbb{C}} \frac{d^2\beta}{\pi} \langle \beta|\alpha \rangle \langle \alpha|A|\beta \rangle = \int_{\mathbb{C}} \frac{d^2\beta}{\pi} \langle \beta|A|\beta \rangle, \quad (4.83)$$

The propagator (4.70) admits a representation in terms of coherent states, which is the following:

$$G(z_f, z_i; t) = \pi^{-n} \int d^2z_{k_1} \ldots d^2z_{k_n} \prod_{j=1}^{n+1} \langle z_j|e^{-iH\varepsilon}|z_{j-1} \rangle. \tag{4.84}$$

We use the notation

$$H(\alpha^*, \beta) := \frac{\langle \alpha|H|\beta \rangle}{\langle \alpha|\beta \rangle}, \tag{4.85}$$

for the expectation value of the Hamiltonian H between the coherent states $|\alpha \rangle$ and $|\beta \rangle$. Then, at order ε, (4.84) reads

$$G(z_f, z_i; t) = \pi^{-n} \int d^2z_{k_1} \ldots d^2z_{k_n} \prod_{j=1}^{n+1} \langle z_j|z_{j-1} \rangle (1 - i\varepsilon H(z_j^*, z_{j-1})). \tag{4.86}$$

For any two arbitrary coherent states, we have the following relation:

$$\langle \alpha|\beta \rangle = \exp\left\{ -\frac{1}{2}|\alpha|^2 - \frac{1}{2}|\beta|^2 + \alpha^*\beta \right\} = \exp\left\{ -\frac{1}{2}(\alpha^*(\alpha - \beta) - (\alpha^* - \beta^*)\beta \right\}. \tag{4.87}$$

Inserting it in (4.86), we obtain for the propagator

$$G(z_f, z_i; t) = \frac{1}{\pi^n} \int d^2z_{k_1} \ldots d^2z_{k_n}$$

$$\times \exp\left\{ i\varepsilon \sum_{j=1}^{n+1} \left[\frac{i}{2}\left(z_j^* \frac{dz_j}{dt} - \frac{dz_j^*}{dt} z_{j-1} \right) - H(z_j^*, z_{j-1}) \right] \right\}. \tag{4.88}$$

Then if we take $z_{j-1} \longmapsto z_j$ for all intermediate steps, we find that

$$G(z_f, z_i; t) = \lim_{n \to \infty} \frac{1}{\pi^n} \int d^2 z_{k_1} \ldots d^2 z_{k_n} \exp \left\{ i \int_0^\infty d\tau \left[\frac{1}{2} \left(z^* \frac{dz}{d\tau} - \frac{dz^*}{d\tau} z \right) - H(z^*, z) \right] \right\}.$$

$$(4.89)$$

This is an expression of the propagator in terms of coherent states. We shall make use of it in the following chapters.

The Liouville Equation

5

5.1   Introduction: Two Points of View

In the sequel, we follow as closely as possible the presentation by Bogolubov [10] of the role of dynamical variables in the statistical properties of a system, either classical or quantum. Let Γ be a classical phase space with n degrees of freedom. Thus, we have n independent generalized coordinates, $q^1, q^2, \ldots, q^n$, and their conjugate momenta being $p_1, p_2, \ldots, p_n$. Henceforth, we represent a point in the phase space Γ as

$$\Gamma = \Gamma(q^1, q^2, \ldots, q^n; p_1, p_2, \ldots, p_n).$$

We assume that we are dealing with a mechanical system subject to external forces deriving from a potential V. In addition, V may depend on the coordinates $q^1, q^2, \ldots, q^n$ (from now on, we shall denote by q the set of all generalized coordinates) and possible on time t, but not on the velocities $\dot{q}$. Let $H = H(q^1, q^2, \ldots, q^n; p_1, p_2, \ldots, p_n; t)$ be the Hamiltonian that governs the time evolution of the system under consideration. The motion can be now described by a system of $2n$ differential equations. In general, this system is non-linear. These are the famous Hamilton equations:

$$\dot{q}^\alpha = \frac{\partial H}{\partial p_\alpha}, \qquad \dot{p}_\alpha = -\frac{\partial H}{\partial q^\alpha}, \qquad \alpha = 1, 2, \ldots, n. \qquad (5.1)$$

In the sequel, we shall always assume that H has no explicit dependence on time.

By definition an *observable* is a *real* function of the coordinate and momenta,

$$A = A(q^1, q^2, \ldots, q^n; p_1, p_2, \ldots, p_n),$$

on the phase space Γ. A *state* is a distribution function, normalized to unity. As an example, the Dirac delta supported at the point $(q^1, q^2, \ldots, q^n; p_1, p_2, \ldots, p_n)$

O. Civitarese and M. Gadella, *Methods in Statistical Mechanics*, Lecture Notes in Physics 974, https://doi.org/10.1007/978-3-030-53658-9_5

in Γ is a state. Usually, this distribution is identified with the point itself, i.e., the point $(q^1, q^2, \ldots, q^n; p_1, p_2, \ldots, p_n)$ represents a state on phase space. From this point of view, dynamical variables or observables are functions of state. We remark that the variables q are generalized coordinates which are not necessarily Cartesian coordinates. The number of independent generalized coordinates is the number of degrees of freedom [10].

The situation in quantum mechanics is somehow different. In quantum mechanics, a pure state is given by a wave function or, equivalently, as an abstract vector ψ in a Hilbert space. The wave function may depend on variables that can represent different physical entities. The most usual among them are the *coordinate*, *momentum* or *energy*. We say that we are using the coordinate, momentum or energy representation, respectively. In addition, the wave function should depend on time, t. In the coordinate representation, the wave function is written as $\psi(\mathbf{x}, t)$, where the vector $\mathbf{x}$ gives the position of one or more particles in Cartesian coordinates. If the system has N particles, the number of coordinates in three dimensions will be equal to $3N$, since $\mathbf{x} = (\mathbf{x}_1, \mathbf{x}_2, \ldots, \mathbf{x}_N)$. The wave function satisfies a dynamical equation, the Schrödinger equation, which plays a similar role as Hamilton equations in classical mechanics. This is

$$i\hbar \frac{\partial \psi(\mathbf{x}, t)}{\partial t} = H \, \psi(\mathbf{x}, t), \tag{5.2}$$

which gives the evolution of the wave function $\psi(\mathbf{x}, t)$ with time.

Contrary to the current situation in classical mechanics, in quantum mechanics dynamical variables are not, generally speaking, functions of the state (nevertheless, there exist exceptions as the von Neumann entropy). These variables are linear maps, also called operators, on a Hilbert space containing all possible wave functions (or pure states) of the system. As is well known, predictions of the result of the measurement of an observable on a given state are not deterministic, in general, but only probabilistic. Only if the vector state (or wave function) ψ is an eigenvector of the dynamical variable represented by the operator A, i.e., $A\psi = a\psi$, we may affirm that a measurement of A on ψ will give a as a result with certitude. In general, the best we can obtain will be the *mean value* or *average* of A on ψ, which is defined as the inner product $(\psi, A\psi)$ in the case that ψ be normalized to one, i.e., $(\psi, \psi) = 1$.

5.2 Classical Time Evolution

In the following, Ω will represent both a point in phase space $\Omega \equiv (q^1, \ldots, q^n; p_1, \ldots, p_n)$ or the state represented by this point, that is, the Dirac delta density function supported at this point.

Let us assume that the dynamics is governed by the time-independent Hamiltonian H. At initial time equal to t_0 one point in phase space is at $\Omega(t_0)$. After evolution, at time t this point is at $\Omega(t)$. Let us call G_{t,t_0} the transformation that carries $\Omega(t_0)$ into $\Omega(t)$. Then, $G_{t,t_0}(\Omega(t_0)) = \Omega(t)$, so that $\Omega(t_0) = G_{t_0,t_0}\Omega(t_0)$, i.e., G_{t_0,t_0} is the identity on phase space. Furthermore, we assume that G_{t,t_0} is independent on the initial point and that it is distributive, which means that

$$\Omega(t_2) = G_{t_2,t_1}(\Omega(t_1)) = G_{t_2,t_1} G_{t_1,t_0}(\Omega(t_0)),$$
$$\Omega(t_2) = G_{t_2,t_0}(\Omega(t_0)). \tag{5.3}$$

Since this must be true for any initial state Ω, we have that

$$\boxed{G_{t_2,t_0} = G_{t_2,t_1} G_{t_1,t_0}}. \tag{5.4}$$

In particular, if we write $t_2 = t_0$ and $t_1 = t$, we have that for any initial state Ω:

$$\Omega = G_{t_0,t_0}(\Omega) = G_{t_0,t}(G_{t,t_0}(\Omega)). \tag{5.5}$$

As Eq. (5.5) is valid for any t_0 and t, it shows that $G_{t_0,t}$ is invertible and

$$G_{t_0,t} = G_{t,t_0}^{-1} \quad \text{and} \quad G_{t_0,t_0} = 1. \tag{5.6}$$

Note that expressions like $G_{t,t'}$ denote transformations on phase space which depend on the values of time t and t', but not on the point Ω at which they apply.

5.2.1 Distribution Probabilities

Let $\mathcal{D}(\Omega)$ be a measurable[1] function on the phase space Γ, with the following properties:

(i) Positivity[2]: $\mathcal{D}(\Omega) \geq 0$.
(ii) Normalization:

$$\int_\Gamma \mathcal{D}(\Omega)\, d\Omega = 1, \tag{5.7}$$

where $d\Omega = dq^1\, dq^2 \ldots dq^n\, dp_1\, dp_2 \ldots dp_n$ is the Lebesgue measure on the phase space Γ and the integral extends to all Γ.

Now, let us consider the dynamical variable or observable $A = A(\Omega)$ on Γ. Its time evolution is defined with the help of the time evolution on Γ using the following *duality* formula:

$$A(\Omega_t) := A(G_{t,t_0}(\Omega_0)), \tag{5.8}$$

---

[1] This is a technical concept that we are not going to explain here. Along the present text, we consider continuous and pointwise continuous functions only and both types are measurable.
[2] As a matter of fact, this is positivity almost elsewhere, which means that $\mathcal{D}(\Omega)$ may not be positive on a set of zero Lebesgue measure.

where the initial point $\Omega_0 \in \Gamma$ is arbitrary.

We have defined, in general, a state as a normalized distribution function on Γ or its corresponding density function (whenever can be defined as a positive function or generalized function). Then, we define the *mean value* or *average* of $A(\Omega)$ on $\mathcal{D}(\Omega)$ as

$$\langle A \rangle = \int_\Gamma A(\Omega)\,\mathcal{D}(\Omega)\,d\Omega \tag{5.9}$$

and the mean value of $A_t(\Omega)$ on some initial density $\mathcal{D}_0(\Omega_0)$ as

$$\langle A_t \rangle = \int_\Gamma A(G_{t,t_0}(\Omega_0))\,\mathcal{D}_0(\Omega_0)\,d\Omega_0\,. \tag{5.10}$$

We denote the Lebesgue measure on $\mathbb{R}^{2n}$ either as $d\Omega = dq^1\,dq^2\ldots dq^n\,dp_1\,dp_2\ldots dp_n$ or as $d\Omega_0 = dq_0^1\,dq_0^2\ldots dq_0^n\,dp_1^0\,dp_2^0\ldots dp_n^0$. The difference is a simple change of variables.

Next, we define the following density on phase space:

$$\mathcal{D}_{t,t_0}(\Omega) := \int_\Gamma \delta(G_{t,t_0}(\Omega_0) - \Omega)\,\mathcal{D}_0(\Omega_0)\,d\Omega_0\,, \tag{5.11}$$

and write

$$
\begin{aligned}
\int_\Gamma A(\Omega)\,\mathcal{D}_{t,t_0}(\Omega)\,d\Omega &= \int_\Gamma A(\Omega)\,d\Omega \int_\Gamma \delta(G_{t,t_0}(\Omega_0) - \Omega)\,\mathcal{D}_0(\Omega_0)\,d\Omega_0 \\
&= \int_\Gamma \mathcal{D}_0(\Omega_0)\,d\Omega_0 \int_\Gamma A(\Omega)\,\delta(G_{t,t_0}(\Omega_0) - \Omega)\,d\Omega \\
&= \int_\Gamma \mathcal{D}_0(\Omega_0)\,A(G_{t,t_0}(\Omega_0))\,d\Omega_0\,.
\end{aligned}
\tag{5.12}
$$

Since, Ω and Ω_0 are dummy variables, we arrive to the following duality formula for the densities $\mathcal{D}_{t,t_0}$ and $\mathcal{D}_0$:

$$\int_\Gamma A(\Omega)\,\mathcal{D}_{t,t_0}(\Omega)\,d\Omega = \int_\Gamma A(G_{t,t_0}(\Omega))\,\mathcal{D}_0(\Omega)\,d\Omega\,. \tag{5.13}$$

Both Eqs. (5.11) and (5.5) show that

$$\mathcal{D}_{t,t_0}(\Omega) = \mathcal{D}_0(G_{t,t_0}(\Omega))\,. \tag{5.14}$$

Using (5.14) in (5.13), we obtain

$$\int_\Gamma A(\Omega)\,\mathcal{D}_0(G_{t,t_0}(\Omega))\,d\Omega = \int_\Gamma A(G_{t,t_0}(\Omega))\,\mathcal{D}_0(\Omega)\,d\Omega\,, \tag{5.15}$$

which is obviously another form of the duality formula that relates the time evolution of states and observables.[3]

We are going to make use of these expression, in the next subsection, to demonstrate sum important results pertaining to statistical mechanics.

5.2.2   Liouville Equation

In this subsection, we shall demonstrate the Liouville equation. To begin with, we need to calculate the derivative:

$$\frac{\partial}{\partial t} \mathcal{D}_{t,t_0}(\Omega). \tag{5.16}$$

To this end, let us recall that the time evolution on phase space is governed by the Hamiltonian $H = H(q^1, q^2, \ldots, q^n, p_1, p_2, \ldots, p_n)$. Then, if we write

$$\Omega(t) = (q^1(t), q^2(t), \ldots, q^n(t), p_1(t), p_2(t), \ldots, p_n(t)),$$

its time derivative is given by

$$\phi(t, \Omega) := \frac{\partial \Omega}{\partial t} = \left(\frac{\partial q^1(t)}{\partial t}, \frac{\partial q^2(t)}{\partial t}, \ldots, \frac{\partial q^n(t)}{\partial t}, \frac{\partial p_1(t)}{\partial t}, \frac{\partial p_2(t)}{\partial t} \cdots, \frac{\partial p_n(t)}{\partial t} \right)$$

$$= \left(\frac{\partial H}{\partial p_1}, \frac{\partial H}{\partial p_2}, \ldots, \frac{\partial H}{\partial p_n}, -\frac{\partial H}{\partial q^1}, -\frac{\partial H}{\partial q^2}, \ldots, -\frac{\partial H}{\partial q^n} \right), \tag{5.17}$$

where the last identity is a consequence of the canonical Hamilton equations (5.1).

Next, since $\Omega(t) = G_{t,t_0}(\Omega_0)$ in (5.17), we have

$$\frac{\partial \Omega}{\partial t} = \frac{\partial G_{t,t_0}(\Omega_0)}{\partial t} = \phi(t, G_{t,t_0}(\Omega_0)). \tag{5.18}$$

Take now the formal partial derivative with respect to t in (5.11). We get

$$\frac{\partial \mathcal{D}_{t,t_0}(\Omega)}{\partial t} = \int_\Gamma \frac{\partial \, \delta(G_{t,t_0}(\Omega_0) - \Omega)}{\partial (G_{t,t_0}(\Omega_0))} \frac{\partial (G_{t,t_0}(\Omega_0))}{\partial t} \mathcal{D}_0(\Omega_0) \, d\Omega_0$$

$$= \int_\Gamma \phi(t, \Omega_t) \left[\frac{\partial}{\partial \Omega_t} \delta(\Omega_t - \Omega) \right] \mathcal{D}_0(\Omega_0) \, d\Omega_0, \tag{5.19}$$

where $\Omega_t := G_{t,t_0}(\Omega_0)$. Due to the properties of the Dirac delta, we have that

$$\phi(t, \Omega_t) \frac{\partial}{\partial \Omega_t} \delta(\Omega_t - \Omega) = -\frac{\partial}{\partial \Omega} (\phi(t, \Omega_t) \delta(\Omega_t - \Omega)) = -\frac{\partial}{\partial \Omega} (\phi(t, \Omega) \delta(\Omega_t - \Omega)). \tag{5.20}$$

---

[3]Note that this formula is similar to that one giving the equivalence of mean values in quantum mechanics in both Schrödinger and Heisenberg representations, respectively.

The second identity in (5.20) is obvious. The other comes from:

$$-\frac{\partial}{\partial\Omega}\left(\phi(t,\Omega)\,\delta(\Omega_t-\Omega)\right)=-\left[\frac{\partial\phi(t,\Omega)}{\partial\Omega}\right]\delta(\Omega_t-\Omega)+\phi(t,\Omega)\,\frac{\partial}{\partial\Omega_t}\,\delta(\Omega_t-\Omega)$$

$$=-\left[\frac{\partial\phi(t,\Omega)}{\partial\Omega_t}\right]\delta(\Omega_t-\Omega)+\phi(t,\Omega)\,\frac{\partial}{\partial\Omega_t}\,\delta(\Omega_t-\Omega)=\phi(t,\Omega)\,\frac{\partial}{\partial\Omega_t}\,\delta(\Omega_t-\Omega),\quad(5.21)$$

where we have used the properties of the Dirac delta and that the vector $\phi(t,\Omega)$ does not depend on Ω_t, explicitly. Then, from (5.19)–(5.21), it follows that:

$$\frac{\partial\mathcal{D}_{t,t_0}(\Omega)}{\partial t}=-\frac{\partial}{\partial\Omega}\int_\Gamma\phi(t,\Omega)\,\mathcal{D}_0(\Omega_0)\,\delta(\Omega_t-\Omega)\,d\Omega_0$$

$$=-\frac{\partial}{\partial\Omega}\left[\phi(t,\Omega)\int_\Gamma\mathcal{D}_0(\Omega_0)\,\delta(G_{t,t_0}(\Omega_0)-\Omega)\,d\Omega_0\right]=-\frac{\partial}{\partial\Omega}\left(\phi(t,\Omega)\,\mathcal{D}_{t,t_0}(\Omega)\right)$$

$$=-\sum_{j=1}^n\left(\frac{\partial}{\partial q^j}\left[\phi_j(t,\Omega)\,\mathcal{D}_{t,t_0}(\Omega)\right]+\frac{\partial}{\partial p_j}\left[\phi_{j+n}(t,\Omega)\,\mathcal{D}_{t,t_0}(\Omega)\right]\right).\quad(5.22)$$

A few remarks on (5.22). Since $\phi(t,\Omega)$ is a vector with $2n$ components, the expression in the right-hand side of the second row in (5.22) is clearly a gradient. In the third row of (5.22), $\phi_k(t,\Omega)$ is the kth component of $\phi(t,\Omega)$. Inserting (5.17) in (5.22) gives

$$\frac{\partial}{\partial t}\,\mathcal{D}_{t,t_0}(\Omega)=-\sum_{j=1}^n\left(\frac{\partial}{\partial q^j}\left[\frac{\partial H}{\partial p_j}\,\mathcal{D}_{t,t_0}(\Omega)\right]-\frac{\partial}{\partial p_j}\left[\frac{\partial H}{\partial q^j}\,\mathcal{D}_{t,t_0}(\Omega)\right]\right)$$

$$=-\sum_{j=1}^n\left(\frac{\partial H}{\partial p_j}\frac{\partial\mathcal{D}_{t,t_0}(\Omega)}{\partial q^j}-\frac{\partial H}{\partial q^j}\frac{\partial\mathcal{D}_{t,t_0}(\Omega)}{\partial p_j}\right).\quad(5.23)$$

As in previous chapters, we use the convention of summation over repeated indices to write

$$\frac{\partial}{\partial t}\,\mathcal{D}_{t,t_0}(\Omega)=\frac{\partial H}{\partial q^\alpha}\frac{\partial\mathcal{D}_{t,t_0}(\Omega)}{\partial p_\alpha}-\frac{\partial H}{\partial p_\alpha}\frac{\partial\mathcal{D}_{t,t_0}(\Omega)}{\partial q^\alpha}.\quad(5.24)$$

At this point, let us recall that if A and B are functions of the coordinates and momenta, Ω, their *Poisson bracket* is defined as

$$\{A,B\}:=\frac{\partial A(\Omega)}{\partial q^\alpha}\frac{\partial B(\Omega)}{\partial p_\alpha}-\frac{\partial A(\Omega)}{\partial p_\alpha}\frac{\partial B(\Omega)}{\partial q^\alpha}.\quad(5.25)$$

The use of Poisson brackets allows for a further simplification of Eq. (5.22), which can be finally written as

$$\frac{\partial}{\partial t}\,\mathcal{D}_{t,t_0}(\Omega)=\{H,\mathcal{D}_{t,t_0}(\Omega)\}.\quad(5.26)$$

Equation (5.26) is the celebrated *Liouville equation*. It gives the evolution equation of any state, defined as a probability density on phase space, under the following conditions:

(i) The movement of a given point Ω in phase space is governed by a Hamiltonian H through the canonical Hamilton equations.

(ii) The Hamiltonian H may depend on the coordinates and momenta, although it may not depend explicitly on time t.

When the motion in phase space is given by a Hamiltonian, it is usually called *Hamiltonian flow*. However, not all motions in phase space are produced by a Hamiltonian. For instance, let us consider the following equation of motion on a two-dimensional phase space:

$$\dot{q} = qp \quad ; \quad \dot{p} = -qp ,$$

where the dot means derivative with respect to time. This system is solvable and its general solution, being given the initial conditions q_0 for q and p_0 for p, is given by

$$q(t) = q_0 \frac{(q_0 + p_0) e^{(q_0 + p_0) t}}{p_0 + q_0 e^{(q_0 + p_0) t}} \quad ; \quad p(t) = p_0 \frac{q_0 + p_0}{p_0 + q_0 e^{(q_0 + p_0) t}} .$$

However, this evolution is not a Hamiltonian flow. For if a Hamiltonian H existed such that

$$\dot{q} = qp = \frac{\partial H}{\partial p} , \qquad \dot{p} = -qp = -\frac{\partial H}{\partial q} ,$$

it should satisfy that

$$\frac{\partial^2 H}{\partial q \partial p} = \frac{\partial^2 H}{\partial p \partial q} .$$

However, as one readily sees from the above equations:

$$\frac{\partial^2 H}{\partial q \partial p} = p , \qquad \frac{\partial^2 H}{\partial p \partial q} = q ,$$

so that H could not exist. This is an example of a flow on phase space which is not Hamiltonian. On the other hand, a flow in phase space is Hamiltonian if and only if for any pair of functions $A(\Omega)$ and $B(\Omega)$, one has that

$$\frac{d}{dt} \{A, B\} = \{\dot{A}, B\} + \{A, \dot{B}\} .$$

5.2.3 A Consequence of the Liouville Equation

Let us go back to Eq. (5.11) and note that if $\mathcal{D}_0(\Omega_0)$ is a constant, then so is $\mathcal{D}_{t,t_0}(\Omega)$. A probability density cannot be a constant on the *whole* phase space, unless that this phase space be compact. Then, we shall always assume that $\mathcal{D}_0(\Omega_0)$ is constant on a subset of phase space and zero otherwise. After (5.11), this implies that $\mathcal{D}_{t,t_0}(\Omega)$

should be constant on a certain subset of Γ and zero outside this set. Using this result in (5.13), we have

$$\int_{\Gamma} A(\Omega)\, d\Omega = \int_{\Gamma} A(G_{t,t_0}(\Omega))\, d\Omega\,, \qquad (5.27)$$

for any integrable function $A(\Omega)$ on the phase space.

Then, let us consider a bounded subset Θ of Γ ($\Theta \subset \Gamma$) at time t, and let $A(\Omega)$ be the characteristic function of Θ:

$$A(\Omega) = \begin{cases} 1 \text{ if } \Omega \in \Theta \\ 0 \text{ if } \Omega \notin \Theta \end{cases}. \qquad (5.28)$$

At a time t', this subset has evolved to be $G_{t',t}(\Theta)$. Then, as an obvious consequence of (5.5), we have

$$\begin{cases} A(G_{t',t}(\Omega)) = 1 \text{ if } \Omega \in G_{t',t}(\Theta) \\ A(G_{t',t}(\Omega)) = 0 \text{ if } \Omega \notin G_{t',t}(\Theta) \end{cases}, \qquad (5.29)$$

which shows that

$$\int_{G_{t',t}(\Theta)} d\Omega = \int_{\Theta} d\Omega\,, \qquad d\Omega = \prod_{i} dq_i\, dp^i\,, \qquad (5.30)$$

or, equivalently,

$$\int_{G_{t',t}(\Theta)} d\Omega = \int_{\Theta} d\Omega\,. \qquad (5.31)$$

Let us recall our assumption that our system is Hamiltonian, so that the evolution with time of any point in phase space is governed by the canonical Hamilton equations. Formulas (5.30) and (5.31) show that under this assumption, *volumes are conserved with time on phase space.*

5.2.4   The Liouvillian

Take a class X of square-integrable functions on phase space. Choose for this class the Hilbert space $L^2(\mathbb{R}^{2n})$. This choice has clear advantages, like the use of a scalar product or a proper definition for transformations of functions as linear operators.

For each function $f(\Omega)$ in X, the time evolution is given by the action of the operator S_{t,t_0} such that

$$S_{t,t_0} f(\Omega) := f(G_{t,t_0}(\Omega))\,. \qquad (5.32)$$

From the properties of G_{t,t_0}, one can easily derive the properties of S_{t,t_0}. In particular:

1. Distributivity: Let t, t' and t'' be three values of time. Then,

$$S_{t'',t'} S_{t',t} f(\Omega) = S_{t'',t'} f(G_{t',t}(\Omega)) = f(G_{t'',t'}(G_{t',t}(\Omega))) = f(G_{t'',t} f(\Omega)) = S_{t'',t} f(\Omega), \quad (5.33)$$

which shows that

$$S_{t'',t'} S_{t',t} = S_{t'',t}. \quad (5.34)$$

2. Invertibility:

$$S_{t,t'} S_{t',t} f(\Omega) = S_{t,t'} f(G_{t',t}(\Omega)) = f(G_{t,t'} G_{t',t}(\Omega)) = f(G_{t,t}(\Omega)) = f(\Omega), \quad (5.35)$$

which implies that

$$S_{t,t'} S_{t',t} = I, \quad (5.36)$$

where I is the identity transformation, i.e., $I f(\Omega) = f(\Omega)$. From here, this is a simple exercise to show that both $S_{t,t'}$ and $S_{t',t}$ are invertible and

$$S_{t,t'}^{-1} = S_{t',t}. \quad (5.37)$$

From (5.37) and (5.32), we obtain the following relation:

$$f(\Omega) = S_{t_0,t} f(G_{t,t_0}(\Omega)). \quad (5.38)$$

Formula (5.32) is applicable for any density function or state and this, in particular, means that

$$\mathcal{D}_0(G_{t,t_0}(\Omega)) = S_{t,t_0} \mathcal{D}_0(\Omega). \quad (5.39)$$

Then, (5.39) together with (5.14) gives

$$\mathcal{D}_{t,t_0}(\Omega) = S_{t,t_0} \mathcal{D}_0(\Omega). \quad (5.40)$$

In order to define the *Liouvillian*, $\mathcal{L}_t$, associated to a given Hamiltonian H, one needs to go back to Eq. (5.26). If we write this equation as

$$\frac{\partial}{\partial t} \mathcal{D}_{t,t_0}(\Omega) = -\mathcal{L}_t \mathcal{D}_{t,t_0}(\Omega), \quad (5.41)$$

one sees by comparison to (5.26) that

$$\mathcal{L} := \frac{\partial H}{\partial p_\alpha} \frac{\partial}{\partial q^\alpha} - \frac{\partial H}{\partial q^\alpha} \frac{\partial}{\partial p_\alpha}, \quad (5.42)$$

where again we are summing over repeated Greek indices. Here, $\alpha = 1, 2, \ldots, n$, where n is the number of degrees of freedom. We can also define the action of the Liouvillian on an arbitrary function $f(\Omega)$ on phase space, which according to (5.42) is

$$\mathcal{L} f(\Omega) = \{H, f\}, \tag{5.43}$$

where $\{-.-\}$ denotes Poisson brackets.

As we have mentioned before, the set of admissible functions belongs to a class called X. Now, we want to be more precise and shall assume that X is the Hilbert space of complex square-integrable functions, in the sense of Lebesgue, on the real line $\mathbb{R}$. Thus, $X \equiv L^2(\mathbb{R}^{2n})$. As is well known, if $f(\Omega)$ and $g(\Omega)$ are two arbitrary functions in $L^2(\mathbb{R}^{2n})$, their scalar product is given by

$$(f, g) := \int_{-\infty}^{\infty} f^*(\Omega) \, g(\Omega) \, d\Omega, \tag{5.44}$$

where the asterisk denotes complex conjugation. This is valid for the quantum case. For the classical case, the use of real Hilbert spaces would have been sufficient, i.e., spaces of real functions over the field of real numbers. In any case, we can consider the Liouvillian $\mathcal{L}$ as an unbounded operator on $L^2(\mathbb{R}^{2n})$. Note that the Liouvillian as an operator on $L^2(\mathbb{R}^{2n})$ is a derivation, and therefore it cannot be defined for any function on $L^2(\mathbb{R}^{2n})$. Its domain is the space of absolutely continuous functions with derivatives in $L^2(\mathbb{R}^{2n})$.

Before to proceed with our discussion, we need a few definitions. Here, we shall consider operators A on $L^2(\mathbb{R}^{2n})$, which may or may not be defined on all vectors of $L^2(\mathbb{R}^{2n})$.

Definition 5.1 The *domain* $D(A)$ of the operator A is the subspace of $L^2(\mathbb{R}^{2n})$ of all the functions on which A acts. Only for bounded operators this domain is the whole Hilbert space. In any case, $D(A)$ must be dense in $L^2(\mathbb{R}^{2n})$, in order to guarantee the existence of the adjoint.

Definition 5.2 An operator A with domain $D(A)$ is called Hermitian if for any pair of functions $f(\Omega)$ and $g(\Omega)$ in $D(A)$, we have that

$$(Af, g) = (f, Ag), \tag{5.45}$$

where the scalar product $(-, -)$ is given in (5.44).

Definition 5.3 An operator A with domain $D(A)$ is called anti-Hermitian if for any pair of functions $f(\Omega)$ and $g(\Omega)$ in $D(A)$, we have that

$$(Af, g) = -(f, Ag). \tag{5.46}$$

In the sequel, we are going to prove that the Liouville operator $\mathcal{L}$ is anti-Hermitian provided that its domain is properly chosen. To this end, let us write

$$(f, \mathcal{L} g) = \int_\Gamma f^*(\Omega) \, (\mathcal{L} g)(\Omega) \, d\Omega = \int_\Gamma f^*(\Omega) \left[\frac{\partial H}{\partial p_\alpha} \frac{\partial g}{\partial q^\alpha} - \frac{\partial H}{\partial q^\alpha} \frac{\partial g}{\partial p_\alpha} \right] d\Omega. \quad (5.47)$$

After these simple and obvious operations (always sum over repeated Greek indices):

$$f^* \frac{\partial H}{\partial p_\alpha} \frac{\partial g}{\partial q^\alpha} = \frac{\partial}{\partial q^\alpha} \left[\frac{\partial H}{\partial p_\alpha} f^* g \right] - \left(\frac{\partial^2 H}{\partial q^\alpha \partial p_\alpha} \right) f^* g - \frac{\partial H}{\partial p_\alpha} \frac{\partial f^*}{\partial q^\alpha} g, \quad (5.48)$$

$$f^* \frac{\partial H}{\partial q^\alpha} \frac{\partial g}{\partial p_\alpha} = \frac{\partial}{\partial p_\alpha} \left[\frac{\partial H}{\partial q^\alpha} f^* g \right] - \left(\frac{\partial^2 H}{\partial q^\alpha \partial p_\alpha} \right) f^* g - \frac{\partial H}{\partial q^\alpha} \frac{\partial f^*}{\partial p_\alpha} g, \quad (5.49)$$

we can insert (5.48) and (5.49) in the right-hand side of (5.47), so as to obtain

$$(f, \mathcal{L} g) = \int_\Gamma \left\{ \frac{\partial}{\partial q^\alpha} \left[\frac{\partial H}{\partial p_\alpha} f^* g \right] - \frac{\partial}{\partial p_\alpha} \left[\frac{\partial H}{\partial q^\alpha} f^* g \right] \right\} d\Omega$$

$$- \int_\Gamma \left[\frac{\partial H}{\partial p_\alpha} \frac{\partial f^*}{\partial q^\alpha} - \frac{\partial H}{\partial q^\alpha} \frac{\partial f^*}{\partial p_\alpha} \right] g(\Omega) \, d\Omega$$

$$= \int_\Gamma \left\{ \frac{\partial}{\partial q^\alpha} \left[\frac{\partial H}{\partial p_\alpha} f^* g \right] - \frac{\partial}{\partial p_\alpha} \left[\frac{\partial H}{\partial q^\alpha} f^* g \right] \right\} d\Omega - (\mathcal{L} f, g). \quad (5.50)$$

We have now to choose a proper domain for $\mathcal{L}$ and this can be the space of functions admitting partial derivatives at all orders and such that they and all their partial derivatives go to zero at the infinity faster than the inverse of any polynomial. This is the Schwartz space $\mathcal{S}(\mathbb{R}^{2n})$. This space is dense in $L^2(\mathbb{R}^{2n})$. One could have use another space of functions vanishing at the infinity, but the Schwartz space is particularly convenient due to the functional form that H may have. Then, for any pair of functions $f(\Omega)$ and $g(\Omega)$ in this domain, the term

$$\int_\Gamma \left\{ \frac{\partial}{\partial q^\alpha} \left[\frac{\partial H}{\partial p_\alpha} f^* g \right] - \frac{\partial}{\partial p_\alpha} \left[\frac{\partial H}{\partial q^\alpha} f^* g \right] \right\} d\Omega \quad (5.51)'$$

vanishes. This is essentially an integration by parts. To illustrate it, let us assume that we have one degree of freedom only. Let us use the notation:

$$w(q, p) = \frac{\partial H(q, p)}{\partial p} f^*(q, p) g(q, p).$$

Then, the integral of the first term in (5.51) becomes (provided that $w(q, p)$ be smooth, so that the order of integration be irrelevant)

$$\int_\Gamma \frac{\partial w(q, p)}{\partial q} dq \, dp = \int_{-\infty}^\infty dp \int_{-\infty}^\infty dq \, \frac{\partial w(q, p)}{\partial q} = \int_{-\infty}^\infty dp \left[\frac{\partial w(q, p)}{\partial q} \Big|_{q=-\infty}^{q=+\infty} \right] = 0,$$

provided that $w(q, p)$ vanishes at the infinity. Consequently,

$$(f, \mathcal{L} g) = -(\mathcal{L} f, g), \tag{5.52}$$

and therefore we conclude that the Liouville operator (also called the Liouvillian) is anti-Hermitian. This property is often described as $\mathcal{L}^\dagger = -\mathcal{L}$, where $\mathcal{L}^\dagger$ is the adjoint of $\mathcal{L}$. However, this is not correct, in principle. One has to take into account that the domains of both operators may be different and, in fact, the domain of the adjoint should be bigger. Such an identity is only possible if both domains were identical. In general, we should write $\mathcal{L} \prec -\mathcal{L}^\dagger$, meaning that minus the adjoint of the Liouvillian extends the Liouvillian.

We shall discuss the role of the Liouville operator in quantum statistical mechanics later in this chapter. At this point, we shall limit ourselves to discuss some properties concerning the Liouville operator as was defined here. For instance, let us assume the existence of a linear operator Λ on $L^2(\Gamma)$, which does not depend on time, such that

(i) The operator Λ commutes with the Liouvillian:

$$\mathcal{L}\Lambda = \Lambda\mathcal{L}. \tag{5.53}$$

(ii) There exists some state $\mathcal{D}_0$ on the phase space Γ such that $\mathcal{D}_0$ is an eigenfunction of Λ with eigenvalue λ:

$$\Lambda\mathcal{D}_0 = \lambda\mathcal{D}_0. \tag{5.54}$$

If at time t_0 the state of a given system is $\mathcal{D}_0$, then at another time t, the state will be $\mathcal{D}_{t,t_0}$, as given in (5.40). Needless to say that $\mathcal{D}_{t,t_0}$ satisfies the Liouville equation (5.41). Since Λ commutes with $\mathcal{L}$, we have (recall that Λ is time independent)

$$\frac{\partial}{\partial t} (\Lambda\, \mathcal{D}_{t,t_0}) = \Lambda \frac{\partial}{\partial t} \mathcal{D}_{t,t_0} = -\Lambda\mathcal{L}\, \mathcal{D}_{t,t_0} = -\mathcal{L}(\Lambda\, \mathcal{D}_{t,t_0}). \tag{5.55}$$

Combining (5.40) and (5.55), we obtain

$$\frac{\partial}{\partial t} (\Lambda\, \mathcal{D}_{t,t_0} - \lambda\mathcal{D}_{t,t_0}) = -\mathcal{L}(\Lambda\, \mathcal{D}_{t,t_0} - \lambda\mathcal{D}_{t,t_0}). \tag{5.56}$$

If $F(t) := \Lambda\, \mathcal{D}_{t,t_0} - \lambda\mathcal{D}_{t,t_0}$, Eq. (5.55) has the following *formal* solution:

$$F(t) = e^{-\int_{t_0}^{t} \mathcal{L}\, dt}\, F(t_0). \tag{5.57}$$

After (5.54), the initial condition is given by $F(t_0) = 0$, so that $F(t) = 0$ for all values of t, which means that

$$\Lambda\, \mathcal{D}_{t,t_0} = \lambda\mathcal{D}_{t,t_0}, \tag{5.58}$$

for all values of t. Therefore, under the previously stated conditions, if at a given time a state is an eigenstate of a given operator, it would keep being an eigenstate for all values of time with the same eigenvalue.

This result has applications in the determination of symmetry properties of $\mathcal{D}_{t,t_0}$. For instance, let us consider a dynamical system formed by N identical particles in three dimensions. In absence of constraints, this system has $3N$ coordinates and $3N$ momenta. Let us denote by σ an arbitrary permutation of $3N$ digits and denote by P_σ the operator which produces the reordering of coordinates and momenta:

$$P_\sigma \mathbf{q} = P(q_1, q_2, \ldots, q_{3N}) := (q_{\sigma(1)}, q_{\sigma(2)}, \ldots, q_{\sigma(3N)}),$$
$$P_\sigma \mathbf{p} = P(p_1, p_2, \ldots, p_{3N}) := (p_{\sigma(1)}, p_{\sigma(2)}, \ldots, p_{\sigma(3N)}). \tag{5.59}$$

Let $f(\mathbf{q}, \mathbf{p})$ be any function on the $6N$-dimensional phase space and Λ be an operator acting on the vector space of these functions such that

$$\Lambda f(\mathbf{q}, \mathbf{p}) := f(P_\sigma \mathbf{q}, P_\sigma \mathbf{p}). \tag{5.60}$$

Since the particles are identical and are not subject to constraints, physical properties will not change under a permutation of positions and momenta. In particular, the Hamiltonian will be invariant under these permutations:

$$\Lambda H(\mathbf{q}, \mathbf{p}) = H(P_\sigma \mathbf{q}, P_\sigma \mathbf{p}) = H(\mathbf{q}, \mathbf{p}). \tag{5.61}$$

Then, we can prove the following property:

For any differentiable function $f(\mathbf{q}, \mathbf{p})$, one has the following identity:

$$\{\Lambda H(\mathbf{q}, \mathbf{p}), f(\mathbf{q}, \mathbf{p})\} = \{H(P_\sigma \mathbf{q}, P_\sigma \mathbf{p}), f(\mathbf{q}, \mathbf{p})\}$$

$$= \{H(\mathbf{q}, \mathbf{p}), f(P_\sigma \mathbf{q}, P_\sigma \mathbf{p})\} = \{H(\mathbf{q}, \mathbf{p}), \Lambda f(\mathbf{q}, \mathbf{p})\}, \tag{5.62}$$

where $\{-, -\}$ denotes Poisson brackets. Let us illustrate the proof for the case of two degrees of freedom. Note first that for any function $f(\mathbf{q}, \mathbf{p})$ on the four-dimensional phase space we have that $\Lambda f(q^1, q^2, p_1, p_2) = f(q^2, q^1, p_2, p_1)$. Then,

$$\{\Lambda H(\mathbf{q}, \mathbf{p}), f(\mathbf{q}, \mathbf{p})\} = \frac{\partial H}{\partial q^2}\frac{\partial f}{\partial p_1} + \frac{\partial H}{\partial q^1}\frac{\partial f}{\partial p_2} - \frac{\partial H}{\partial p_2}\frac{\partial f}{\partial q_1} - \frac{\partial H}{\partial p_1}\frac{\partial f}{\partial q^2}, \tag{5.63}$$

and on the other side, we have

$$\{H(\mathbf{q}, \mathbf{p}), \Lambda f(\mathbf{q}, \mathbf{p})\} = \frac{\partial H}{\partial q^1}\frac{\partial f}{\partial p_2} + \frac{\partial H}{\partial q^2}\frac{\partial f}{\partial p_1} - \frac{\partial H}{\partial p_1}\frac{\partial f}{\partial q^2} - \frac{\partial H}{\partial p_2}\frac{\partial f}{\partial q^1}. \tag{5.64}$$

We see that Eqs. (5.63) and (5.64) are identical. The proof in the general case is based in this idea and it is left to the reader.

Now, if we define $(\Lambda \mathcal{L}) f(\mathbf{q}, \mathbf{p}) := \{\Lambda H, f\}$, the previous result shows that for any function $f(\mathbf{q}, \mathbf{p})$, one has

$$(\mathcal{L}_t \Lambda) f(\mathbf{q}, \mathbf{p}) = (\Lambda \mathcal{L}_t) f(\mathbf{q}, \mathbf{p}) \Longrightarrow \mathcal{L}_t \Lambda = \Lambda \mathcal{L}_t . \tag{5.65}$$

Observe that, according to this definition, we have

$$(\Lambda \mathcal{L}) = \frac{\partial H(P_\sigma \mathbf{q}, P_\sigma \mathbf{p})}{\partial q^\alpha} \frac{\partial}{\partial p_\alpha} - \frac{\partial H(P_\sigma \mathbf{q}, P_\sigma \mathbf{p})}{\partial p_\alpha} \frac{\partial}{\partial q^\alpha} ,$$

where we assume summation over repeated Greek indices.

Next, let us assume that the function $\mathcal{D}_0(\mathbf{q}, \mathbf{p})$, representing a state, is symmetric with respect to its degrees of freedom, i.e.,

$$\Lambda \mathcal{D}_0(\mathbf{q}, \mathbf{p}) = \mathcal{D}_0(P\mathbf{q}, P\mathbf{p}) = \mathcal{D}_0(\mathbf{q}, \mathbf{p}) , \tag{5.66}$$

so that $\mathcal{D}_0(\mathbf{q}, \mathbf{p})$ is an eigenstate of Λ with eigenvalue $\lambda = 1$. Then, due to (5.65) and (5.58), we conclude that after a time t this symmetry is preserved so that

$$\Lambda \mathcal{D}_{t,t_0}(\mathbf{q}, \mathbf{p}) = \mathcal{D}_{t,t_0}(P_\sigma \mathbf{q}, P_\sigma \mathbf{p}) = \mathcal{D}_{t,t_0}(\mathbf{q}, \mathbf{p}) . \tag{5.67}$$

It is interesting to note that the invariance of the Hamiltonian with respect to *some* of the degrees of freedom would produce the same symmetry properties restricted to these degrees of freedom.

Equation (5.57) gives for any function $F(t, \mathbf{q}(t), \mathbf{p}(t))$ the following time evolution:

$$F(t, \mathbf{q}(t), \mathbf{p}(t)) = e^{-(t-t_0)\mathcal{L}} F(t_0, \mathbf{q}(t_0), \mathbf{p}(t_0)) . \tag{5.68}$$

In particular, if the value of the function $F(t, \mathbf{q}(t), \mathbf{p}(t))$ at $t = t_0$ is $\mathcal{D}_0$, we have

$$\mathcal{D}_{t,t_0} = e^{-(t-t_0)\mathcal{L}} \mathcal{D}_0 = S_{t,t_0} \mathcal{D}_0 . \tag{5.69}$$

The second identity in (5.69) is a consequence of (5.40). Since the initial state $\mathcal{D}_0$ is arbitrary, we have the following identity between operators, valid for all values of t and t_0:

$$S_{t,t_0} = e^{-(t-t_0)\mathcal{L}} . \tag{5.70}$$

Henceforth, if we take $t_0 = 0$, it is convenient to use the notation $S_t := e^{-t\mathcal{L}}$ and $\mathcal{D}_t = \mathcal{D}_{t,0}$, so that one can write

$$\mathcal{D}_t = S_t \mathcal{D}_0 = e^{-t\mathcal{L}} \mathcal{D}_0 . \tag{5.71}$$

We are going to discuss now an interesting property which concerns to time evolution of observables. The mean value of an observable $A(\Omega)$ on a state $\mathcal{D}_0(\Omega)$ was given in (5.9). Now, let us assume that we have a time evolution governed by the time-independent Hamiltonian H, or equivalently by its Liouvillian $\mathcal{L}$. According to (5.71), if at $t = 0$ the state is given by $\mathcal{D}_0(\Omega)$ at time t it becomes $\mathcal{D}_t$. At time t, the mean value of $A(\Omega)$ on $\mathcal{D}_t$ is

$$\langle A \rangle_t = \int_\Gamma A(\Omega)\, \mathcal{D}_t(\Omega)\, d\Omega = \int_\Gamma A_t(\Omega)\, \mathcal{D}_0(\Omega)\, d\Omega\,, \qquad (5.72)$$

where the second identity in (5.72) *defines* the evolution of the observable $A(\Omega)$ with time. In order to obtain an equation equivalent to (5.71) for observables, let us proceed with the following *formal* manipulations:

$$\langle A \rangle_t = \int_\Gamma A(\Omega) e^{-t\mathcal{L}} \mathcal{D}_0(\Omega)\, d\Omega = \int_\Gamma A(\Omega) \left[\sum_{n=0}^\infty \frac{(-1)^n t^n}{n!} \mathcal{L}^n \mathcal{D}_0(\Omega) \right] d\Omega. \quad (5.73)$$

Due to (5.52), we have[4]

$$\int_\Gamma A(\Omega)\, [\mathcal{L}\, \mathcal{D}_0(\Omega)]\, d\Omega = - \int_\Gamma [\mathcal{L}\, A(\Omega)]\, \mathcal{D}_0(\Omega)\, d\Omega\,, \qquad (5.74)$$

so that

$$\int_\Gamma A(\Omega)\, [\mathcal{L}^n\, \mathcal{D}_0(\Omega)]\, d\Omega = (-1)^n \int_\Gamma [\mathcal{L}^n\, A(\Omega)]\, \mathcal{D}_0(\Omega)\, d\Omega\,. \qquad (5.75)$$

Then, using (5.73) in (5.75), we obtain

$$\langle A \rangle_t = \int_\Gamma \left[\sum_{n=0}^\infty \frac{t^n}{n!} \mathcal{L}^n\, A(\Omega) \right] \mathcal{D}_0(\Omega)\, d\Omega$$

$$= \int_\Gamma \left[e^{t\mathcal{L}}\, A(\Omega) \right] \mathcal{D}_0(\Omega)\, d\Omega = \int_\Gamma A_t(\Omega)\, \mathcal{D}_0(\Omega)\, d\Omega\,. \qquad (5.76)$$

Since the initial state $\mathcal{D}_0(\Omega)$ is arbitrary, we conclude that

$$A_t(\Omega) = e^{t\mathcal{L}}\, A(\Omega) = S_t\, A(\Omega)\,. \qquad (5.77)$$

The formal partial derivative with respect to time in (5.77) gives

$$\frac{\partial A_t}{\partial t} = \mathcal{L}\, A_t \iff \frac{\partial A_t}{\partial t} = \{ A_t, H \}\,, \qquad (5.78)$$

which gives the time evolution for any observable $A(\Omega)$ on the phase space Γ.

---

[4]We have assumed in principle that $A(\Omega)$ and $\mathcal{D}_0(\Omega)$ are square-integrable functions, although formulas like (5.52) can be extended to more general situations.

5.3   Liouville Equations in Quantum Mechanics

To begin with, we are using a notation permitting the study of systems with the maximum generality. Thus, our variable $\mathbf{x}$ not only denotes position or other kind of continuous coordinate, but instead serves for any kind of coordinates, either continuous or discrete, necessary to describe the properties of the system. If we have a system of s different types of N_a identical particles, with $a = 1, 2, \ldots, s$, then

$$\mathbf{x} \equiv (\mathbf{x}_{1,1}, \ldots, \mathbf{x}_{j,a}, \ldots, \mathbf{x}_{N_a,s}), \tag{5.79}$$

where $\mathbf{x}_{j,a}$ represents the coordinates of the jth particle of type ath. Here, $j = 1, 2, \ldots, N_a$ and $a = 1, 2, \ldots, s$. Each of the generalized coordinates $\mathbf{x}_{j,a}$ contains all the degrees of freedom of the $\{j, a\}$th particle, so that

$$\mathbf{x}_{j,a} = (\mathbf{q}_{j,a}, \boldsymbol{\sigma}_{j,a}), \tag{5.80}$$

where $\boldsymbol{\sigma}_{j,a}$ stands for internal degrees of freedom such as spin, etc. In this compact notation, the wave function of the system is denoted as $\phi(\mathbf{x})$. The scalar product of two wave functions is

$$(\phi_1, \phi_2) = \int \phi_1^*(\mathbf{x})\phi_2(\mathbf{x})\,d\mathbf{x}, \tag{5.81}$$

where the star denotes complex conjugation. The measure is $d\mathbf{x} = d\mathbf{x}_{1,1} \ldots d\mathbf{x}_{N_a,s}$, where each $d\mathbf{x}_{j,a}$ can be decomposed into a continuous and a discrete part as in (5.80), so that

$$\int \ldots d\mathbf{x}_{j,a} = \sum_{\mathbf{v}_{ja}} \int \ldots d\mathbf{q}_{j,a}. \tag{5.82}$$

Operators are linear transformations on the space of wave functions. A common form for an operator A when acting on a wave function is represented by

$$(A\phi)(\mathbf{x}) := \int A(\mathbf{x}, \mathbf{x}')\,\phi(\mathbf{x}')\,d\mathbf{x}'. \tag{5.83}$$

The time evolution of the wave function is given by the time-dependent Schrödinger equation. Calling to $\phi(t, \mathbf{x})$ the time-dependent wave function, the mean value of the operator A at the time t is

$$\langle A \rangle_t = \int \phi^*(t, \mathbf{x}) A(\mathbf{x}, \mathbf{x}')\,\phi(t, \mathbf{x}')\,d\mathbf{x}\,d\mathbf{x}'. \tag{5.84}$$

Let us assume that $\psi_1(t, \mathbf{x}), \psi_2(t, \mathbf{x}), \ldots, \psi_n(t, \mathbf{x}), \ldots$ is a sequence of normalized wave function solutions of the Schrödinger equation at a given time $t = t_0$. The functions in this sequence may or may not form a complete orthonormal basis. In this, generally speaking, subspace each wave function $\phi(t, \mathbf{x})$ may be written as

$$\phi(t, \mathbf{x}) = \sum_n c_n \psi_n(t, \mathbf{x}). \tag{5.85}$$

Let $\{\omega_n\}$ be a sequence of positive numbers ($\omega_n \geq 0$) such that

$$\sum_n \omega_n = 1. \tag{5.86}$$

We define the *density matrix* or *statistical average* of the quantum states $\{\psi_n(t, \mathbf{x})\}$ with weights $\omega_n \geq 0$ as

$$\mathcal{D}_t(\mathbf{x}, \mathbf{x}') := \sum_n \omega_n \psi_n^*(t, \mathbf{x}) \, \psi_n(t, \mathbf{x}'). \tag{5.87}$$

The weight ω_n is the probability of the function $\psi_n(t, \mathbf{x})$ in the above average. Expression (5.87) is valid in the basis where the ω_n are the eigenvalues of the matrix $M(c_\alpha^\dagger, c_\beta)$. This is the matrix which represents the tensor product $\phi(t, \mathbf{x}) \otimes \phi(t, \mathbf{x})$, where $\phi(t, \mathbf{x})$ is as in (5.85).

Then, we can easily generalize the mean value (5.84) so as to include statistical averages like (5.87) in the following form:

$$\langle A \rangle_t = \int A(\mathbf{x}, \mathbf{x}') \, \mathcal{D}_t(\mathbf{x}, \mathbf{x}') \, d\mathbf{x} \, d\mathbf{x}', \tag{5.88}$$

which can be written as

$$\langle A \rangle_t = \mathrm{Tr}[A\mathcal{D}_t]. \tag{5.89}$$

We define the action of the operator $\mathcal{D}_t$ as follows: for any function $R(\mathbf{x})$, we have

$$(\mathcal{D}_t \, R)(\mathbf{x}) = \int \mathcal{D}_t(\mathbf{x}, \mathbf{x}') \, R(\mathbf{x}') \, d\mathbf{x}'. \tag{5.90}$$

Although the functions $\psi_n(\mathbf{x})$ need not to be mutually orthogonal, if these functions span the whole Hilbert space, we always can find an orthonormal basis $\{\phi_j(\mathbf{x})\}$ so that

$$\mathcal{D}_t(\mathbf{x}, \mathbf{x}') := \sum_j \lambda_j \eta_j^*(t, \mathbf{x}) \eta_j(t, \mathbf{x}'). \tag{5.91}$$

The weights $\{\omega_n\}$ and $\{\lambda_j\}$ are different in general.

5.3.1 General Properties of the Statistical Averages

These first properties are immediate consequences of definition (5.87):

1. Conjugation under inversion of arguments: $\mathcal{D}_t^*(\mathbf{x}, \mathbf{x}') = \mathcal{D}_t(\mathbf{x}', \mathbf{x})$.
2. Hermiticity: $\mathcal{D}_t^{\dagger} = \mathcal{D}_t$, where the dagger stands for adjoint. In fact, for any square-integrable function (in general complex) $R(x)$, we can write

$$
(R, \mathcal{D}_t^{\dagger} R) = \int R^*(\mathbf{x}) \mathcal{D}_t^*(\mathbf{x}, \mathbf{x}') \, R(\mathbf{x}') \, d\mathbf{x} \, d\mathbf{x}'
$$

$$
= \int R^*(\mathbf{x}) \sum_n \omega_n \, \psi_n(t, \mathbf{x}') \psi_n^*(t, \mathbf{x}) \, R(\mathbf{x}') \, d\mathbf{x} \, d\mathbf{x}'
$$

$$
= \int R^*(\mathbf{x}) \sum_n \omega_n \, \psi_n(t, \mathbf{x}') \psi_n^*(t, \mathbf{x}) \, R(\mathbf{x}') \, d\mathbf{x}' \, d\mathbf{x} = (R, \mathcal{D}_t R) \,. \quad (5.92)
$$

3. Trace one: $\operatorname{Tr} \mathcal{D}_t = 1$.
4. Positivity: Let $R(\mathbf{x})$ be a complex square-integrable function. Then, the scalar product,

$$
(R, \mathcal{D}_t R) = \int R^*(\mathbf{x}) D_t(\mathbf{x}, \mathbf{x}') R(\mathbf{x}') \, d\mathbf{x} \, d\mathbf{x}'
$$

$$
= \int R^*(\mathbf{x}) \sum_n \omega_n \, \psi_n^*(t, t_0, \mathbf{x}) \psi_n(t, t_0, \mathbf{x}') \, R(\mathbf{x}') \, d\mathbf{x} \, d\mathbf{x}'
$$

$$
= \sum_n \omega_n \int R^*(\mathbf{x}) \, \psi_n(t, t_0, \mathbf{x}') \, d\mathbf{x} \int R(\mathbf{x}') \, \psi_n^*(t, t_0, \mathbf{x}) \, d\mathbf{x}'
$$

$$
= \sum_n \omega_n \left| \int R(\mathbf{x}) \, \psi_n(t, \mathbf{x}) \, d\mathbf{x} \right|^2 \geq 0 \,, \quad (5.93)
$$

is positive defined. Positivity is often denoted by

$$
\mathcal{D}_{t, t_0} \geq 0 \,. \quad (5.94)
$$

Note that Hermiticity comes after positivity for bounded operators.

5. We have that $\mathcal{D}_t^2 \leq \mathcal{D}_t$.
6. The operator $\mathcal{D}_t$ is bounded.

5.3.2 Time Evolution of the Statistical Averages

To begin with, let us justify Eq. (5.88). Let $\mathcal{H}$ be the Hilbert space vector represented by the function $\phi(t, \mathbf{x})$. If H is the Hamiltonian governing a time evolution, we may write the average $\langle \phi | H\phi \rangle$ into two different forms:

$$\langle \phi | H\phi \rangle = \int \phi^*(t, \mathbf{x})\, d\mathbf{x}' \int H(\mathbf{x}', \mathbf{x})\, \phi(t, \mathbf{x}')\, d\mathbf{x}, \qquad (5.95)$$

$$\langle H\phi | \phi \rangle = \int H^*(\mathbf{x}, \mathbf{x}')\, \phi^*(t, \mathbf{x})\, d\mathbf{x} \int \phi(t, \mathbf{x}')\, d\mathbf{x}'. \qquad (5.96)$$

Since the Hamiltonian is Hermitian, $\langle \phi | H\phi \rangle = \langle H\phi | \phi \rangle$, then (5.88) holds, i.e., $H(\mathbf{x}', \mathbf{x}) = H^*(\mathbf{x}, \mathbf{x}')$. Note that for full generality, we are admitting that the integral kernel of the Hamiltonian $H(\mathbf{x}', \mathbf{x})$ may be complex. For the arbitrary state $\psi(t, \mathbf{x})$ and its complex conjugate, the Schrödinger equation reads

$$i\hbar \frac{\partial \phi(t, \mathbf{x})}{\partial t} = \int H(\mathbf{x}, \mathbf{x}'')\phi(t, \mathbf{x}'')\, d\mathbf{x}'', \qquad (5.97)$$

$$i\hbar \frac{\partial \phi^*(t, \mathbf{x}')}{\partial t} = -\left[i\hbar \frac{\partial \phi(t, \mathbf{x}')}{\partial t} \right]^* = -\int H^*(\mathbf{x}', \mathbf{x}'')\phi^*(t, \mathbf{x}'')\, d\mathbf{x}''$$

$$= -\int H(t, \mathbf{x}'', \mathbf{x}')\phi^*(t, \mathbf{x}'')\, d\mathbf{x}''. \qquad (5.98)$$

Next, take (5.87) and derive it with respect to t, using (5.97) and (5.98):

$$i\hbar \frac{\partial \mathcal{D}_t(\mathbf{x}, \mathbf{x}')}{\partial t} = i\hbar \frac{\partial}{\partial t} \sum_n \omega_n \psi_n^*(t, \mathbf{x}')\psi_n(t, \mathbf{x})$$

$$- \sum_n \omega_n \int H(\mathbf{x}'', \mathbf{x}')\, \psi_n^*(t, \mathbf{x}'')\psi_n(t, \mathbf{x})\, d\mathbf{x}''$$

$$+ \sum_n \omega_n \int H(\mathbf{x}, \mathbf{x}'')\, \psi_n(t, \mathbf{x}'')\, \psi_n^*(t, \mathbf{x}')\, d\mathbf{x}''$$

$$= \int H(\mathbf{x}, \mathbf{x}'')\, \mathcal{D}_t(\mathbf{x}'', \mathbf{x}')\, d\mathbf{x}'' - \int H(\mathbf{x}'', \mathbf{x}')\, \mathcal{D}_t(\mathbf{x}, \mathbf{x}'')\, d\mathbf{x}''. \qquad (5.99)$$

Then, it is a simple exercise to show that if we have two operators A and B with respective integral kernels $A(\mathbf{x}, \mathbf{x}'')$ and $B(\mathbf{x}'', \mathbf{x})$, the integral kernel for the product is given by

$$(AB)(\mathbf{x}, \mathbf{x}') = \int A(\mathbf{x}, \mathbf{x}'')\, B(\mathbf{x}'', \mathbf{x}')\, d\mathbf{x}''. \qquad (5.100)$$

Note that $(AB)(\mathbf{x}, \mathbf{x}')$ and $(BA)(\mathbf{x}, \mathbf{x}')$ are not necessarily identical, so that the operators may not commute.

Equation (5.99) is a relation between the integral kernels of the Hamiltonian and the operator $\mathcal{D}_t$. This relation gives an equation between the operators, which after (5.100) is written as

$$i\hbar \frac{\partial}{\partial t} \mathcal{D}_t = H\mathcal{D}_t - \mathcal{D}_t H = [H, \mathcal{D}_t], \qquad (5.101)$$

or in a more compact form:

$$\frac{\partial}{\partial t}\,\mathcal{D}_t = \frac{1}{i\hbar}\,[H, \mathcal{D}_t]\,. \tag{5.102}$$

This is the celebrated *Liouville equation*.

5.3.3 The Unitary Operator U_{t,t_0}

By definition, the operator U_{t,t_0} relates the wave function at time $t = t_0$ with the wave function at time t, i.e.,

$$\psi(t, \mathbf{x}) = U_{t,t_0}\psi(t_0, \mathbf{x})\,. \tag{5.103}$$

Then,

$$i\hbar\,\frac{\partial}{\partial t}\,\phi(t, \mathbf{x}) = i\hbar\,\frac{\partial}{\partial t}\,U_{t,t_0}\,\phi(t_0, \mathbf{x})\,,$$

$$i\hbar\,\frac{\partial}{\partial t}\,\phi(t, \mathbf{x}) = H\phi(t, \mathbf{x}) = HU_{t,t_0}\,\phi(t_0, \mathbf{x})\,. \tag{5.104}$$

If we omit the arbitrary initial wave function $\phi(t_0, \mathbf{x})$, we obtain a simple expression:

$$i\hbar\,\frac{\partial}{\partial t}\,U_{t,t_0} = HU_{t,t_0}\,. \tag{5.105}$$

The operator U_{t,t_0} has the following properties that can be deduced straightforwardly after the definition:

1. $U_{t_0,t_0} = I$, where I is the identity operator: $I\phi(t_0, \mathbf{x}) = \phi(t_0, \mathbf{x})$, for any initial wave function.
2. $U_{t,t_0}U_{t_0,t} = U_{t_0,t}U_{t,t_0} = I$.
3. From the previous property, we find that $U_{t,t'} = U_{t',t}^{-1}$, where t and t' are two arbitrary values of time.
4. The adjoint operator obeys the equation of motion

$$-i\hbar\,\frac{\partial}{\partial t}\,U_{t,t_0}^{\dagger} = U_{t,t_0}^{\dagger}H\,. \tag{5.106}$$

5. From (5.106), one obtains

$$i\hbar\,\frac{\partial}{\partial t}(U_{t,t_0}^{\dagger}U_{t,t_0}) = -U_{t,t_0}^{\dagger}HU_{t,t_0} + U_{t,t_0}^{\dagger}HU_{t,t_0} = 0\,. \tag{5.107}$$

We conclude that $U_{t,t_0}^\dagger U_{t,t_0}$ is time independent for any value of t. This property steam from the fact that for any time t_0 it holds $U_{t_0,t_0}^\dagger U_{t_0,t_0} = I$. Therefore,

$$U_{t,t_0}^\dagger U_{t,t_0} = U_{t,t_0} U_{t,t_0}^\dagger = I. \tag{5.108}$$

6. Taken the derivative with respect to time:

$$i\hbar \frac{\partial}{\partial t} (U_{t,t_0}^\dagger \mathcal{D}_t U_{t,t_0})$$

$$= i\hbar \left(\frac{\partial}{\partial t} U_{t,t_0}^\dagger \right) \mathcal{D}_t U_{t,t_0} + i\hbar U_{t,t_0}^\dagger \left(\frac{\partial}{\partial t} \mathcal{D}_t \right) U_{t,t_0} + i\hbar U_{t,t_0}^\dagger \mathcal{D}_t \left(\frac{\partial}{\partial t} U_{t,t_0} \right)$$

$$= -U_{t,t_0}^\dagger H \mathcal{D}_t U_{t,t_0} + U_{t,t_0}^\dagger (H \mathcal{D}_t - \mathcal{D}_t H) U_{t,t_0} + U_{t,t_0}^\dagger \mathcal{D}_t H U_{t,t_0} = 0, \tag{5.109}$$

from where we conclude that

$$\mathcal{D}_0 := U_{t,t_0}^\dagger \mathcal{D}_t U_{t,t_0} \tag{5.110}$$

does not depend on time. $\mathcal{D}_0$ satisfies all properties of the operators $\mathcal{D}_t$. Consequently, it is also a density operator, since by inverting (5.110)

$$\mathcal{D}_t = U_{t,t_0} \mathcal{D}_0 U_{t,t_0}^\dagger. \tag{5.111}$$

Let us go back to Eq. (5.88) that gives the time evolution of the average of the observable A on the state $\mathcal{D}_t$. We recall that it can be written as indicated by the first identity in (5.112)

$$\langle A \rangle_t = \mathrm{Tr}\{A\mathcal{D}_t\} = \mathrm{Tr}\{A U_{t,t_0} \mathcal{D}_0 U_{t,t_0}^\dagger\} = \mathrm{Tr}\{U_{t,t_0}^\dagger A U_{t,t_0} \mathcal{D}_0\} = \mathrm{Tr}\{A_{t,t_0} \mathcal{D}_0\}. \tag{5.112}$$

The second identity in (5.112) is obvious. The third one comes from the invariance of a trace under circular permutations. The last one comes after the definition:

$$A_{t,t_0} = U_{t,t_0}^\dagger A U_{t,t_0} = U_{t_0,t} A U_{t_0,t}^\dagger. \tag{5.113}$$

5.3.4   Some Comments on $\mathcal{D}_0$

All vectors in the basis $\{\psi_n\}$ in the Hilbert space $\mathcal{H}$ of square-integrable functions are eigenvectors of $\mathcal{D}_0$, i.e., $\mathcal{D}_0 \psi_n = \omega_n \psi_n$ and the eigenvalues ω_n are real non-negative numbers. If $\psi_n(\mathbf{x})$ is the wave function that represents the vector ψ_n, the eigenvalue equation can be written in the form:

$$\int \mathcal{D}_0(\mathbf{x}'', \mathbf{x}') \psi_n^*(\mathbf{x}'') \, d\mathbf{x}'' = \omega_n \psi_n^*(\mathbf{x}'). \tag{5.114}$$

As a straightforward consequence of the spectral theorem (see (5.129)), we have

$$\mathcal{D}_0(\mathbf{x}, \mathbf{x}') = \sum_n \omega_n \psi_n^*(\mathbf{x}') \psi_n(\mathbf{x}) . \tag{5.115}$$

5.4   Symmetry Properties

The state functions for systems of fermions or bosons do not exhibit the same symmetry properties, as a consequence of the fulfillment of the Pauli exclusion principle and the same is true for the density operators.

In quantum mechanics, bosons are particles with either zero or integer spin. Fermions have half-integer spin. In quantum statistical mechanics, we shall mainly consider ensembles formed either by bosons or by fermions. Then, if ϕ is a wave function of an ensemble of N identical fermions or bosons, σ an arbitrary permutation of these particles and P_σ its corresponding operator acting on wave functions, as defined in (5.59), we have that

$$P_\sigma \phi = n(P_\sigma)\phi, \quad \text{with} \quad n(P_\sigma) = \begin{cases} 1 & \text{for bosons} \\ (-1)^\sigma = \pm 1 & \text{for fermions} \end{cases} . \tag{5.116}$$

Finally, we are going to see how permutations of particles transform statistical averages (i.e., integral kernels) for density operators. Take a density operator $\mathcal{D}_t$ as in (5.90) and consider the following expression valid for an arbitrary wave function $\phi(\mathbf{x})$:

$$\eta(P_\sigma)[\mathcal{D}_t \phi(\mathbf{x})] = (\mathcal{D}_t P_\sigma^{-1}\phi)(\mathbf{x}) = \int \mathcal{D}_{t,t_0}(\mathbf{x}, \sigma\mathbf{x}') \, \phi(\mathbf{x}') \, d\mathbf{x}' , \tag{5.117}$$

where $\eta(P_\sigma)$ has been defined in (5.116). In the first identity in (5.117), we have taken into account that the particle permutation σ and its inverse σ^{-1} have the same parity so that $\eta(P_\sigma) = \eta(P_\sigma^{-1})$. Then, (5.117) shows the following relation between the statistical averages:

$$\mathcal{D}_t(\mathbf{x}, \sigma\mathbf{x}') = \eta(P_\sigma) \, \mathcal{D}_t(\mathbf{x}, \mathbf{x}') . \tag{5.118}$$

Then, we apply P_σ to the integral expression for $\mathcal{D}_t$,

$$P_\sigma [\mathcal{D}_t \phi](\mathbf{x}) = [\mathcal{D}_t \phi](P\mathbf{x}) = \int \mathcal{D}_t(\sigma\mathbf{x}, \mathbf{x}')\phi(\mathbf{x}') \, d\mathbf{x}' . \tag{5.119}$$

Next, we assume that $\mathcal{D}_t$ is a mixture of pure states all with a well-defined parity. Therefore,

$$P_\sigma \, \mathcal{D}_t P_\sigma^{-1} = P_\sigma \sum_n \omega_n \, |\psi_n\rangle\langle\psi_n| \, P_\sigma^{-1} = \sum_n \omega_n \, |\psi_n\rangle\langle\psi_n| = \mathcal{D}_t , \tag{5.120}$$

since $[\eta(P_\sigma)]^2 = 1$.

Using the language of statistical averages (or integral kernels), we reach to the following expression:

$$\mathcal{D}_t(\mathbf{x}, \mathbf{x}') = \mathcal{D}_t(\sigma\mathbf{x}, \sigma\mathbf{x}'), \tag{5.121}$$

and

$$\mathcal{D}_t(\sigma\mathbf{x}, \mathbf{x}') = \eta(P_\sigma)\,\mathcal{D}_t(\mathbf{x}, \mathbf{x}'). \tag{5.122}$$

5.5 Mathematical Properties of the Density Operator

In this section, we give the main properties of the density operators for the quantum mixture states. The point of view is rather mathematical and is addressed to those with a basic knowledge of the theory of operators on Hilbert spaces.

Definition A density operator ρ on a Hilbert space $\mathcal{H}$ is a linear[5] operator on $\mathcal{H}$ with the following properties:

(i) ρ is a bounded operator on $\mathcal{H}$ and, therefore defined on the whole $\mathcal{H}$, so that no questions about domains are relevant here.
(ii) ρ is a positive operator, $\rho \geq 0$, which means that for any vector $\psi \in \mathcal{H}$, one has $(\psi, \rho\psi) \geq 0$, where $(-, -)$ is the scalar product on $\mathcal{H}$.
(iii) ρ is a trace-class operator. This means that for any orthonormal basis $\{\psi_n\}$ on $\mathcal{H}$, we have that the sum,

$$\mathrm{Tr}\,\rho := \sum_n (\psi_n, \rho\psi_n) < \infty, \tag{5.123}$$

converges. The value of the sum in (5.123) does not depend on the choice of the orthonormal basis $\{\psi_n\}$.

As a consequence of (i) and (ii), any density operator is self-adjoint. A very important result concerning density operators comes from a spectral representation theorem for normal (and therefore self-adjoint) trace-class operators. Before introducing this result, some comments are in order.

Let ψ be any vector on $\mathcal{H}$. Let us denote the one-dimensional subspace spanned by ψ as $[\psi] := \{\varphi \in \mathcal{H} \ ; \ \varphi = \lambda\psi\}$, for some $\lambda \in \mathbb{C}$, where $\mathbb{C}$ is the field of complex numbers. It is well known that $\mathcal{H} = [\psi] \oplus [\psi]^\perp$, where $\oplus$ stands for direct orthogonal sum. This means that any vector $\phi \in \mathcal{H}$ can be written as $\phi = \lambda\psi + \varphi$, where

---

[5]In the sequel, all operators are linear except otherwise stated. In consequence, we henceforth omit the word linear when referring to linear operators.

$(\psi, \varphi) = 0$. We may assume that ψ is normalized to one, i.e., $||\psi|| := \sqrt{(\psi, \psi)} = 1$. The (orthogonal) projection associated to ψ is an operator P_ψ on $\mathcal{H}$, which acts as $P_\psi \phi = \lambda \psi$ for all $\phi \in \mathcal{H}$.

Now, let $\{\psi_1, \psi_2, \ldots, \psi_n, \ldots\}$ be an orthonormal basis on $\mathcal{H}$ and $\{P_1, P_2, \ldots, P_n, \ldots\}$ the sequence of associated projections as above. This sequence has some interesting properties as, for instance, $P_i P_j = \delta_{ij} P_j$, where δ_{ij} is the usual Kronecker delta, or for any $\phi \in \mathcal{H}$, $\sum_i P_i \psi = \psi$, so that $\sum_i P_i = I$, the identity operator. If the dimension of $\mathcal{H}$ is infinite, this sum is valid in the strong convergence sense.

The projections P_i are usually denoted as $P_i = |\psi_i\rangle\langle\psi_i|$.

Then, we are in the position of stating the following result:

- Let ρ be a density operator on a Hilbert space $\mathcal{H}$. Then, there exists an orthonormal basis $\{\psi_i\}$ on $\mathcal{H}$ and a sequence of positive numbers $\{\lambda_i\}$, $\lambda_i \geq 0$, such that

$$\rho = \sum_i \lambda_i P_i = \sum_i \lambda_i |\psi_i\rangle\langle\psi_i|, \tag{5.124}$$

where $\{P_i\} \equiv |\psi_i\rangle\langle\psi_i|$ is the sequence of projections associated to the basis $\{\psi_i\}$.

From this theorem, one may conclude that $\rho\psi_i = \lambda_i \psi_i$ and the trace of ρ is

$$\mathrm{Tr}\, \rho = \sum_i \lambda_i = 1, \tag{5.125}$$

so that the sum of eigenvalues of ρ converges to one. It can be proved that the spectrum of ρ is purely discrete and that the quantities λ_i are the eigenvalues of ρ.

The usual interpretation of ρ as a quantum state is the following: The state ρ is a statistical mixture of the pure states $\{\psi_i\}$. It means that we have an indetermination on which state the system is. It may be in the state ψ_n with a probability equal to λ_n.

If the vectors ψ_i in (5.124) are mutually orthogonal as in the previous theorem, they are unique, except for multiplication by a phase $e^{i\eta}$. If we drop the orthogonality condition, we also drop uniqueness. To show it, let us consider a mixture of the form $\rho := \sum_j \mu_j |\phi_j\rangle\langle\phi_j|$. Here, $\mu \geq 0$ and $\sum_i \mu_i = 1$ also, although the ϕ_i are not necessarily mutually orthogonal. The operator ρ has the following properties:

1. It is bounded. Although a bit technical, this property is not difficult to prove. Let φ be an arbitrary vector on $\mathcal{H}$. Then, since we assume that ϕ_i are pure states and therefore $||\phi_i|| = 1$, one has

$$||\rho\varphi|| = ||\sum_i \mu_i(\phi_i, \varphi)\phi_i|| \leq \sum_{i} \mu_i |(\phi_i, \varphi)| \, ||\phi_i|| \leq \left(\sum_i \mu_i\right) ||\varphi|| = ||\varphi||,$$

where we have used the Schwartz inequality: $|(\phi_i, \varphi)| \leq ||\phi_i|| \, ||\varphi|| = ||\varphi||$.

2. It is positive. For any $\varphi \in \mathcal{H}$, we have

$$(\varphi, \rho\varphi) = \left(\varphi, \left(\sum_j \mu_j |\phi_j\rangle\langle\phi_j|\right)\varphi\right) = \sum_i \mu_i |(\varphi, \phi_i)|^2 \geq 0 . \quad (5.126)$$

3. It has trace one. To show it, let us take an arbitrary orthonormal basis $\{\psi_n\}$ on $\mathcal{H}$. The trace of ρ is then

$$\text{Tr}\,\rho = \sum_n (\psi_n, \rho\psi_n) = \sum_n \left(\psi_n, \left(\sum_j \mu_j |\phi_j\rangle\langle\phi_j|\right)\psi_n\right) = \sum_i \mu_i \sum_n |(\psi_n, \phi_i)|^2$$

$$= \sum_i \mu_i \|\phi_i\|^2 = \sum_i \mu_i = 1 . \quad (5.127)$$

The first equality in the second row in (5.127) is a consequence of the Parseval identity and the next one comes after the normalization of ϕ_i ($\|\phi_i\| = 1$).

The conclusion is that $\rho = \sum_j \mu_j |\phi_j\rangle\langle\phi_j|$ is a density operator. Then, the spectral theorem above applies, so that ρ admits a decomposition of the form (5.124). Since the vectors ψ_i are mutually orthogonal and this is not necessarily the case with the vectors ϕ_i, we conclude that both decompositions for the same density operator are different. Also the weights μ_i and λ_i should not coincide.

4. $\rho^2 \leq \rho$. This means that the operator $\rho - \rho^2 \geq 0$, i.e., it is a positive one. To show it, note that if we write ρ in the form (5.124), we have that

$$\rho - \rho^2 = \sum_n (\lambda_n - \lambda_n^2) |\psi_n\rangle\langle\psi_n| , \quad (5.128)$$

Since $\lambda \leq 1$, we have that $\lambda_n \geq \lambda_n^2$. Therefore, all eigenvalues of $\rho - \rho^2$ are non-negative. Since $\rho - \rho^2$ has purely discrete spectrum, this means that this operator is positive. Note that ρ is a pure state if and only if there is one $\lambda_n = 1$.

Next, we want to find the connection between the statistical averages defined in (5.87) and the density matrix ρ. Take $\rho = \sum_n \omega_n |\psi_n\rangle\langle\psi_n|$ and assume that the vectors ψ_n admit a representation as functions of the form $\psi_n(t, t_0, \mathbf{x})$ as above. Note that if $\psi \equiv \psi(t, t_0, \mathbf{x})$ is an arbitrary function in the Hilbert space of states, we have

$$\langle\psi|\rho\psi\rangle = \sum_n \omega_n \langle\psi|\psi_n\rangle\langle\psi_n|\psi\rangle$$

$$= \sum_n \omega_n \int \psi^*(t, \mathbf{x})\psi_n(t, \mathbf{x})\,d\mathbf{x} \int \psi_n^*(t, \mathbf{x}')\psi(t, \mathbf{x}')\,d\mathbf{x}'$$

$$= \int \psi^*(t, \mathbf{x})\left[\sum_n \omega_n \psi_n^*(t, \mathbf{x}')\,\psi_n(t, \mathbf{x})\right]\psi(t, \mathbf{x}')\,d\mathbf{x}\,d\mathbf{x}'$$

$$= \int \psi^*(t, \mathbf{x})\,\mathcal{D}_t(\mathbf{x}, \mathbf{x}')\,\psi(t, \mathbf{x}')\,d\mathbf{x}\,d\mathbf{x}' , \quad (5.129)$$

so that $\mathcal{D}_t(\mathbf{x}', \mathbf{x})$ is the kernel of the operator ρ. Thus, observe that in our notation $\rho \equiv \mathcal{D}_t$. In consequence, the properties of $\mathcal{D}_t$ are the properties of ρ stated above. In particular, $\mathrm{Tr}\, \mathcal{D}_t = 1$, $\mathcal{D}_t^2 \leq \mathcal{D}_t$ and $\mathcal{D}_t$ is a bounded operator.

Next, let us give a simple derivation of the *Liouville equation* based on the above general formalism. The Schrödinger equation says that

$$i\hbar \frac{\partial \phi(t)}{\partial t} = H\phi(t) \iff i\hbar \frac{\partial |\phi(t)\rangle}{\partial t} = H|\phi(t)\rangle\,, \tag{5.130}$$

where the expression on the right denotes the Schrödinger equation in the *ket* notation. Since we have expressed the density operator ρ in terms of Dirac notation, it is desirable to write the Schrödinger equation in terms of vectors. This is shown in any elementary textbook on quantum mechanics; nevertheless, we give it in here for completeness. Let ψ any state independent of time t. Then

$$i\hbar \frac{\partial}{\partial t} \langle \phi(t)|\psi\rangle = \langle -i\hbar \frac{\partial}{\partial t} \phi(t)|\psi\rangle = \langle -H\phi(t)|\psi\rangle = -\langle H\phi(t)|\psi\rangle$$

$$= -\langle \phi(t)|H\psi\rangle \iff i\hbar \frac{\partial}{\partial t} \langle \phi(t)| = -\langle \phi(t)|H\,. \tag{5.131}$$

Then, the time evolution of ρ is given by

$$i\hbar \frac{\partial}{\partial t} \rho = \sum_n \omega_n \left[i\hbar \frac{\partial}{\partial t} |\psi_n\rangle \right] \langle \psi_n| - \sum_n \omega_n |\psi_n\rangle \left[i\hbar \frac{\partial}{\partial t} \langle \psi_n| \right]$$

$$= \sum_n \omega_n [H|\psi_n\rangle]\langle \psi_n| + \sum_n \omega_n |\psi_n\rangle[\langle \psi_n|H]$$

$$= H\left[\sum_n \omega_n |\psi_n\rangle\langle \psi_n| \right] - \left[\sum_n \omega_n |\psi_n\rangle\langle \psi_n| \right] H = [H, \rho]\,. \tag{5.132}$$

The identity between the first and the last term in (5.132) gives the *Liouville equation*:

$$\boxed{\frac{\partial \rho}{\partial t} = \frac{1}{i\hbar} [H, \rho]\,.} \tag{5.133}$$

Along the present chapter, we have considered trace-class operators defined through an integral kernel of the form:

$$(\rho\phi)(\mathbf{x}) = \int A(\mathbf{x}, \mathbf{x}')\, \phi(\mathbf{x}')\, d\mathbf{x}\, d\mathbf{x}'\,. \tag{5.134}$$

The integral kernel $A(\mathbf{x}, \mathbf{x}')$ has the following important property:

$$\mathrm{Tr}\, \rho = \int |A(\mathbf{x}, \mathbf{x}')|\, d\mathbf{x}\, d\mathbf{x}'\,. \tag{5.135}$$

Canonical Distributions and Thermodynamic Functions

6

In the previous chapters, we have discussed the role of the distribution functions, from the point of view of the Hamiltonian dynamics as well as from the operator approach [1–3]. This chapter is devoted to the unification of these schemes. Also, we shall pay attention to the definition of thermodynamical observables in the context of static and quasi-static processes. We shall begin with the definition of the integrals of motion.

6.1 Integrals of Motion

The equation of motion for the dynamical variable, A, in general time dependent, is given by

$$\frac{dA}{dt} = \frac{\partial A}{\partial t} + [A, H].$$ (6.1)

In the case that the dynamical variable A is a constant of motion, it satisfies the following equation:

$$\frac{dA}{dt} = 0 \implies \frac{\partial A}{\partial t} = [H, A].$$ (6.2)

If furthermore, A does not depend explicitly on time,

$$\frac{\partial A}{\partial t} = 0 \implies [H, A] = 0.$$ (6.3)

© The Editor(s) (if applicable) and The Author(s), under exclusive license
to Springer Nature Switzerland AG 2020
O. Civitarese and M. Gadella, *Methods in Statistical Mechanics*, Lecture Notes
in Physics 974, https://doi.org/10.1007/978-3-030-53658-9_6

Dynamical variables A satisfying (6.3) are called *integrals of motion*. On the other hand, the Liouville theorem gives the following expression for the time evolution of the probability density:

$$\frac{\partial \mathcal{D}}{\partial t} = [H, \mathcal{D}].$$

(6.4)

A probability distribution $\mathcal{D}$ is *stationary*, i.e., it is time invariant if and only if

$$[H, \mathcal{D}] = 0.$$

(6.5)

Then, we say that $\mathcal{D}$ is a *integral of motion*.

6.2   Gibbs Distributions

First of all, we consider the situation within the context of classical mechanics.

Let Σ be a macroscopical system with a *fixed* number of particles, governed by the Hamiltonian H. Assume that Σ is in a state for which the distribution function is the Gibbs distribution function, which is defined as

$$\mathcal{D} := \frac{e^{-\beta H}}{N_0}.$$

(6.6)

Here, N_0 is a normalization constant, which is determined by performing the integral in the domain $\Sigma(q, p)$

$$\int_{\Sigma(q,p)} \mathcal{D} \, d\Omega = 1,$$

(6.7)

we must have that

$$N_0 = \int_{\Sigma(q,p)} e^{-\beta H} \, d\Omega.$$

(6.8)

If the system Σ is divided into subsystems $\Sigma_1, \Sigma_2, \ldots, \Sigma_s$, which either are mutually independent or weakly interacting, the definition of the Gibbs function is generalized to

$$\mathcal{D} = \prod_{i=1}^{s} \frac{1}{N_0(i)} e^{-\beta H_i},$$

(6.9)

where H_i is the Hamiltonian acting on the subsystem Σ_i, $i = 1, 2, \ldots, s$ and $N_0(i)$ is the corresponding normalization constant.

In order to define the notion of *thermodynamic equilibrium* in the sense of Gibbs, we need a slight generalization of the Gibbs state valid for systems with s species of different indistinguishable particles. This definition is assumed valid for all kinds of macroscopical systems under the condition that the number of particles in each of the species remains constant. Also, in order to be tractable, systems under our consideration should be described by N integrals of motion I_j, $j = 1, 2, \ldots, N$. This means that we have to add the Hamiltonian terms of the form $\sum_i \lambda_i I_i$ to enforce the conserved character of the integral of motions. Naturally, the variation of the parameters λ_i guarantees the constancy of the average values of the integrals of motion. Then, in the space of parameters the motion is constrained by the conditions $\sum_i \lambda_i I_i$.

The density function $\mathcal{D}$ remains stationary after the variation of all parameters entering in the definition of the Hamiltonian

$$H \Rightarrow H - \sum_i \lambda_i I_i \,. \tag{6.10}$$

In general, if each Hamiltonian H_i consists of a kinetic energy term and the interaction between the subsystems is given by $\Phi(\mathbf{q}_{j_i} - \mathbf{q}_{j_2})$, then the total Hamiltonian has the following structure:

$$H = \sum_{1 \le j \le N} \left(\frac{\mathbf{p}_j^2}{2m} + \lambda_j I_j \right) + \sum_{1 \le j_1 < j_2 \le N} \Phi(\mathbf{q}_{j_i} - \mathbf{q}_{j_2}) \,. \tag{6.11}$$

The Gibbs function associated to this Hamiltonian has the form:

$$\mathcal{D} = \frac{1}{c} \, W(\mathbf{q}_1, \mathbf{q}_2, \ldots, \mathbf{q}_N) \exp \left\{ -\beta \sum_{1 \le j \le N} \frac{\mathbf{p}_j^2}{2m} \right\} \,, \tag{6.12}$$

where

$$W(\mathbf{q}_1, \mathbf{q}_2, \ldots, \mathbf{q}_N) = \exp \left\{ -\beta \left(\sum_{1 \le j \le N} \lambda_j I_j - \sum_{1 \le j_1 < j_2 \le N} \Phi(\mathbf{q}_{j_i} - \mathbf{q}_{j_2}) \right) \right\} \,. \tag{6.13}$$

6.3   Thermodynamic Construction in Quasi-static Processes

As we have discussed in previous chapters, the free energy F depends on β and H. Therefore, it depends on the parameters fixing H. In addition, F is an extensive variable, so that it must depend on the volume of the system, V. Then, we can write the dependence of F as

$$F \equiv F(\beta, V, \lambda_1, \ldots, \lambda_l) \,. \tag{6.14}$$

Consequently,

$$dF = \frac{\partial F}{\partial \beta} d\beta + \frac{\partial F}{\partial V} dV + \sum_{1 \le j \le l} \frac{\partial F}{\partial \lambda_j} d\lambda_j . \tag{6.15}$$

The different terms of dF are explicitly calculated as follows:

$$\frac{\partial F}{\partial \beta} d\beta = S dT \tag{6.16}$$

and

$$\frac{\partial F}{\partial \lambda_j} = \frac{1}{\text{Tr}\{e^{-\beta H}\}} \frac{\partial}{\partial \lambda_j} \text{Tr}\{e^{-\beta H}\} , \tag{6.17}$$

since

$$\frac{\partial}{\partial \lambda_j} \text{Tr}\{e^{-\beta H}\} = \text{Tr}\left\{\frac{\partial}{\partial \lambda_j}\{e^{-\beta H}\}\right\} = \text{Tr}\left\{-\beta \frac{\partial H}{\partial \lambda_j} e^{-\beta H} , \right\} \tag{6.18}$$

which shows that (6.17) is equal to

$$\frac{\partial F}{\partial \lambda_j} = -\beta \frac{1}{\text{Tr}\{e^{-\beta H}\}} \text{Tr}\left\{\frac{\partial H}{\partial \lambda_j} e^{-\beta H}\right\} \tag{6.19}$$

or

$$\frac{\partial F}{\partial \lambda_j} = -\beta \left\langle \frac{\partial H}{\partial \lambda_j} \right\rangle ; \tag{6.20}$$

where we have used brackets in order to denote the average of the quantity enclosed by them.

It remains to calculate $\frac{\partial F}{\partial V}$. If all coordinates and momenta are scaled by a parameter λ, such that

$$\mathbf{q} \Rightarrow \lambda \mathbf{q}' \quad \text{and} \quad \mathbf{p} \Rightarrow \frac{1}{\lambda} \mathbf{p}' . \tag{6.21}$$

After this change of coordinates, the Hamiltonian becomes a function of the scale λ,

$$H \equiv H(\lambda, \lambda_1, \ldots, \lambda_l) =: H_\lambda . \tag{6.22}$$

The trace is a functional over the algebra of trace-class operators on a Hilbert space. We have made a transformation on phase space that induces a transformation on the algebra of observables of the system. By duality, this transformation induces a new trace that we denote as Tr_λ, which is defined through the following duality form:

$$\mathrm{Tr}\{e^{-\beta H_\lambda}\} = \mathrm{Tr}_\lambda\{e^{-\beta H}\}. \tag{6.23}$$

Since $V_\lambda = \lambda^3 V$, keeping the lowest order in the variation $\delta\lambda$, we get

$$\delta V \approx 3V\,\delta\lambda. \tag{6.24}$$

Replacing the partial derivative with respect to volume by the partial derivative with respect to λ,

$$\frac{\partial}{\partial V} \Rightarrow \frac{\partial}{\partial\lambda}$$

in dF in (6.15), we obtain

$$\frac{\partial}{\partial V}\mathrm{Tr}\{e^{-\beta H}\} = \left[\frac{\partial}{\partial V}\mathrm{Tr}_\lambda\{e^{-\beta H}\}\right]_{\lambda=1} = \frac{1}{3V}\left[\frac{\partial}{\partial\lambda}\mathrm{Tr}\{e^{-\beta H_\lambda}\}\right]_{\lambda=1}$$
$$= -\frac{\beta}{3V}\left[\mathrm{Tr}\left\{\frac{\partial H_\lambda}{\partial\lambda}e^{-\beta H_\lambda}\right\}\right]_{\lambda=1}. \tag{6.25}$$

Then,

$$\frac{\partial F}{\partial V} = \frac{1}{3V}\left\langle\frac{\partial H_\lambda}{\partial\lambda}\right\rangle_{\lambda=1}, \tag{6.26}$$

and finally

$$\frac{\partial F}{\partial V}dV = \left\langle\frac{\partial H_\lambda}{\partial\lambda}\right\rangle_{\lambda=1}d\lambda, \tag{6.27}$$

which gives the second term of (6.15) as an expression including an average of the derivative with respect to λ of the transformed Hamiltonian H_λ. In summary:

$$dF = -S\,dT + \left\langle\frac{\partial H}{\partial\lambda}\right\rangle_{\lambda=1}d\lambda + \sum_{j=1}^{l}\left\langle\frac{\partial H}{\partial\lambda_j}\right\rangle d\lambda_j. \tag{6.28}$$

Formula (6.28) can be rewritten in terms of the pressure P using the equivalence:

$$P = -\frac{1}{3V}\left\langle\frac{\partial H}{\partial V}\right\rangle_{\lambda=1} \Longrightarrow \left\langle\frac{\partial H}{\partial\lambda}\right\rangle_{\lambda=1}d\lambda = -P\,dV. \tag{6.29}$$

In consequence,

$$dF = -S\,dT - P\,dV - \sum_{j=1}^{l} I_j\,d\lambda_j, \tag{6.30}$$

because

$$I_j = -\left\langle \frac{\partial H}{\partial \lambda_j} \right\rangle. \tag{6.31}$$

Now, the internal energy is written as

$$dE = d(F + ST) = -P\,dV - \sum_{j=1}^{l} I_j\,d\lambda_j + T\,dS. \tag{6.32}$$

The physical meaning of terms in (6.32) is the following:

1. dE represents the *change in the internal energy.*

2. $-P\,dV$ is the change of internal energy associated to the change in the volume at constant pressure P.

3. The expression $-\sum_{j=1}^{l} I_j\,d\lambda_j$ gives the change in the internal energy associated to changes in the external parameters λ_j, $j = 1, 2, \dots, l$.

4. $T\,dS$ is the change in the internal energy associated to a change in the entropy at constant temperature.

5. The variations $(2 + 3)$ give the change of energy produced by the work due to external actions, while 4 is the change due to heat transmission.

Next, let us define the *specific heat* or *heat capacity* at *constant volume* as

$$C_V := T\,\frac{\partial}{\partial T}\left(\frac{S}{V}\right). \tag{6.33}$$

Thus, for constant volume and no variation of the external parameters, we have after (6.32) that

$$\frac{\partial E}{\partial T} = T\,\frac{\partial S}{\partial T}. \tag{6.34}$$

Expression (6.34) is one of the famous Maxwell's relation between thermodynamic quantities. From this relation between the partial derivatives of the entropy and the energy, we recover the known expression of the specific heat at constant volume

$$C_V = \frac{T}{V}\left(\frac{1}{T}\,\frac{\partial E}{\partial T}\right) = \frac{\partial}{\partial T}\left(\frac{E}{V}\right). \tag{6.35}$$

6.3.1 Homogeneous Systems

Let us consider a *macroscopic* system, which contains s different subsystems each one formed by different types of identical particles so that the total number of particles N is the sum of the particles of each subsystem, $N = N_1 + N_2 + \cdots + N_s$.

We say that this macroscopic system is *homogeneous* if the density of particles of each type is constant. More precisely, if

$$\lim_{V \to \infty} \frac{N_j}{V} = n_j = \text{constant}, \qquad j = 1, 2, \ldots, s, \tag{6.36}$$

n_j being the density of the jth type. Now the question is whether the following limits exists:

$$\lim_{V \to \infty} \frac{F}{V} =: f(\beta, a_1, \ldots, a_s; n_1, \ldots, n_s) \tag{6.37}$$

$$\lim_{V \to \infty} \frac{E}{V} = \mathcal{E}(\beta, a_1, \ldots, a_s; n_1, \ldots, n_s), \tag{6.38}$$

where $a_1, a_2, \ldots, a_s$ are some parameters associated to the species of particles that made up the system.

If, in addition, for given values of the parameters $(\beta, a_1, \ldots, a_s; n_1, \ldots, n_s)$, we have that

$$F = Vf \quad \text{and} \quad E = V\mathcal{E}, \tag{6.39}$$

then F and E are *extensive* quantities. In general, they must depend on the system volume.

Physical quantities that do not depend on the volume are called *intensive*. Typical examples are the densities f, $\mathcal{E}$ and also the specific heat C_V.

6.3.2 Fluctuations

Let us consider the *fluctuations* characterized by the *second moment* of the Hamiltonian on a given quantum state, which are defined as

$$\langle (H - E)^2 \rangle, \tag{6.40}$$

where $\langle A \rangle$ denotes the mean value of the observable represented by the operator A on the given state. Here $E = \langle H \rangle$ is the average of the Hamiltonian. Let us take the derivative of E with respect to β. It gives

$$\frac{\partial E}{\partial \beta} = \frac{1}{(\text{Tr}\, e^{-\beta H})^2} \left[-\text{Tr}\{H^2 e^{-\beta H}\} \text{Tr}\{e^{-\beta H}\} + (\text{Tr}\{H e^{-\beta H}\})^2 \right]$$

$$= -\frac{\text{Tr}\{H^2 e^{-\beta H}\}}{\text{Tr}\{e^{-\beta H}\}} + \left[\frac{\text{Tr}\{H e^{-\beta H}\}}{\text{Tr}\{e^{-\beta H}\}} \right]^2$$

$$= -\langle H^2 \rangle + \langle H \rangle^2 = -\langle H^2 - E^2 \rangle, \tag{6.41}$$

where the average $\langle - \rangle$ has been taken with respect to the Gibbs state. Note that, due to the relation $E = \langle H \rangle$, we have

$$\langle (H - E)^2 \rangle = \langle H^2 - 2HE + E^2 \rangle = \langle H^2 \rangle - 2E\langle H \rangle + E^2 = \langle H^2 - E^2 \rangle,$$

(6.42)

so that

$$\langle (H - E)^2 \rangle = -\frac{\partial E}{\partial \beta}.$$

(6.43)

Equation (6.43) may be used to write a relation between the second moment $\langle (H - E)^2 \rangle$ and the specific heat C_V,

$$C_V = \beta^2 \frac{k_B}{V} \langle (H - E)^2 \rangle.$$

(6.44)

6.3.3 Representations

Similar expressions can be derived for quantum systems. As before, we assume that the Hamiltonian H has a purely discrete spectrum with eigenvalues E_v and respective eigenvectors $|v\rangle$

$$H|v\rangle = E_v|v\rangle.$$

(6.45)

Let ω_v be the probability that the state $|v\rangle$ with energy E_v be occupied. Note that this means that the density should be given by a mixture of the form:

$$\mathcal{D} = \sum_v \omega_v |v\rangle\langle v|.$$

(6.46)

Since the total probability must be equal to one, we have that

$$\sum_v \omega_v = 1.$$

(6.47)

If the system is in equilibrium, in the sense of Gibbs, then these probabilities are just

$$\omega_v = \frac{e^{-\beta E_v}}{\mathrm{Tr}\{e^{-\beta H}\}} = \frac{e^{-\beta E_v}}{\sum_{v'}\langle v'|e^{-\beta H}|v'\rangle} = \frac{e^{-\beta E_v}}{\sum_{v'} e^{-\beta E_{v'}}}.$$

(6.48)

Then, the second moment $\langle (H - E)^2 \rangle$ is

$$\langle (H - E)^2 \rangle \equiv \mathrm{Tr}\, \mathcal{D}\,(H - E)^2 = \sum_v \omega_v \langle v|(H^2 - 2EH + E^2)|v\rangle =$$

$$= \sum_v \omega_v (E_v - E)^2 = \sum_v \omega_v (E_v - V\mathcal{E})^2 = V^2 \sum_v \omega_v \left(\frac{E_v}{V} - \mathcal{E}\right)^2.$$

(6.49)

Using (6.44), we find

$$C_V = \beta^2 V k_B \sum_v \omega_v \left(\frac{E_v}{V} - \mathcal{E}\right)^2 ,$$

(6.50)

then,

$$\sum_v \omega_v \left(\frac{E_v}{V} - \mathcal{E}\right)^2 = k_B T^2 \left(\frac{C_V}{V}\right) .$$

(6.51)

The quadratic mean fluctuation of the energy given by the second moment $\langle(H - E)^2\rangle$ measures the distribution of the eigenstate energies per volume, with respect to $\mathcal{E}$.

6.4   The Grand Canonical Ensemble

In this section, we are going to consider a quantum system formed by s species of different indistinguishable particles, to which we assign the density distribution,

$$\mathcal{D} := \frac{1}{\mathrm{Tr}\,\{e^{-\beta(H - \sum_{j=1}^s \mu_j M_j)}\}} \, e^{-\beta(H + \sum_{j=1}^s \mu_j N)_j} ,$$

(6.52)

where μ_j, $j = 1, 2, \ldots, s$ are s constants and the N_j, for each $j = 1, 2, \ldots, s$, are the operators that *determine* the number of particles of the species j.

The volume of the system is fixed as V, the spectrum of H is discrete, ϕ_ω are the eigenstates of H, with eigenvalues E_ω. The number of particles of each species, N_j, is a constant of motion. In consequence,

$$[H, N_j] = 0, \qquad j = 1, 2, \ldots, s .$$

(6.53)

Then,

$$H\phi_\omega = E_\omega \phi_\omega , \qquad N_j \phi_\omega = n_j \phi_\omega \qquad j = 1, 2, \ldots, s ,$$

(6.54)

where n_j denotes the number of particles of each species, e.g.,

$$n_j = \langle N_j \rangle = \mathrm{Tr}\,(N_j \mathcal{D}) .$$

(6.55)

In this representation, the free energy is

$$\mathfrak{F} := -\frac{1}{\beta} \ln(\mathrm{Tr}\,e^{-\beta\mathfrak{H}}) ,$$

(6.56)

where

$$\mathfrak{H} := H - \sum_{j=1}^{s} \mu_j N_j \,. \tag{6.57}$$

Note that $\mathfrak{F}$ is a straightforward generalization of F in equation (6.14). After the definition of $\mathfrak{H}$, we note that it commutes with all the number operators and also with H. This fact legitimates the following manipulations: take the derivative of (6.57) with respect to μ_j. This gives

$$\frac{\partial \mathfrak{F}}{\partial \mu_j} = -\frac{1}{\beta} \frac{1}{\mathrm{Tr}\,\{e^{-\beta\mathfrak{H}}\}} \mathrm{Tr} \left(\frac{\partial}{\partial \mu_j} e^{-\beta\mathfrak{H}} \right) = \frac{1}{\mathrm{Tr}\,\{e^{-\beta\mathfrak{H}}\}} \mathrm{Tr} \left(\frac{\partial\mathfrak{H}}{\partial \mu_j} e^{-\beta\mathfrak{H}} \right) . \tag{6.58}$$

Since obviously,

$$\frac{\partial\mathfrak{H}}{\partial \mu_j} = N_j \,, \qquad j = 1, 2, \ldots, s \,, \tag{6.59}$$

relation (6.58) yields to

$$\frac{\partial \mathfrak{F}}{\partial \mu_j} = -\frac{1}{\mathrm{Tr}\,\{e^{-\beta\mathfrak{H}}\}} \mathrm{Tr} \left(N_j \, e^{-\beta\mathfrak{H}} \right) = -\langle N_j \rangle = -n_j \,, \qquad j = 1, 2, \ldots, s \,, \tag{6.60}$$

or

$$\boxed{\frac{\partial \mathfrak{F}}{\partial \mu_j} = -n_j \,, \qquad j = 1, 2, \ldots, s \,,} \tag{6.61}$$

which is valid at constant β.

6.4.1   On the Uniqueness of the Values for μ_j

Here, we want to discuss the uniqueness of the solutions of equations (6.61). To this end, assume that for s given values of the number of particles of each species $(n_1, n_2, \ldots, n_s)$, we have two distinct solutions, for the set of Lagrange multipliers μ^j, this is

$$\mu^0 = (\mu_1^0, \mu_2^0, \ldots, \mu_s^0) \tag{6.62}$$

$$\mu^1 = (\mu_1^1, \mu_2^1, \ldots, \mu_s^1) \,. \tag{6.63}$$

Then, let us consider the variable τ with the condition $0 \le \tau \le 1$, and define the variation

$$\mu(\tau) = \mu^0 + \tau \Delta\mu \,, \qquad \Delta\mu := \mu^1 - \mu^0 \,. \tag{6.64}$$

Observe that the relations in (6.64) could be interpreted as vector relations with s components. Also note that (6.64) implies that

$$\mu(0) = \mu^0, \qquad \mu(1) = \mu^1, \qquad (6.65)$$

which are also relations with s components. With these definitions, we construct the following function:

$$U(\tau) := \mathfrak{F}(\mu(\tau)) + \sum_{j=1}^{s} N_j \mu_j(\tau). \qquad (6.66)$$

Derive (6.66) with respect to the parameter τ to obtain

$$\frac{dU(\tau)}{d\tau} = \sum_{j=1}^{s} \left(\frac{\partial \mathfrak{F}}{\partial \mu_j} \right) \frac{d\mu(\tau)}{d\tau} + N_j \frac{d\mu_j(\tau)}{d\tau}$$

$$= \sum_{j=1}^{s} \left(\frac{\partial \mathfrak{F}}{\partial \mu_j} \right) \Delta u_j + N_j \Delta \mu_j = \sum_{j=1}^{s} \left(\frac{\partial \mathfrak{F}}{\partial \mu_j} + N_j \right) \Delta \mu_j. \qquad (6.67)$$

Here, we have used that $d\mu(\tau) = d\tau \, \Delta\mu$ as a direct consequence of (6.64). Due to (6.61), the parenthesis in the last term in (6.67) vanishes for $\tau = 0$ and $\tau = 1$, which are the values of τ for which we assume that the solution exists. Therefore,

$$\left(\frac{dU(\tau)}{d\tau} \right)_{\tau=0} = \left(\frac{dU(\tau)}{d\tau} \right)_{\tau=1} = 0, \qquad (6.68)$$

so that

$$\int_0^1 \frac{d^2 U(\tau)}{d\tau^2} \, d\tau = 0. \qquad (6.69)$$

In addition, since N_j and $\Delta\mu_j$ for all j do not depend on τ, we have that

$$\frac{d^2 U(\tau)}{d\tau^2} = \sum_j \left(\frac{d}{d\tau} \frac{\partial \mathfrak{F}}{\partial \mu_j} \right) \Delta\mu_j = \sum_{j,j'} \left(\frac{\partial^2 \mathfrak{F}}{\partial \mu_j \mu_{j'}} \right) \Delta\mu_j \, \Delta\mu_{j'}. \qquad (6.70)$$

Next, let us go back to (6.60) that for simplicity we write here as

$$-\frac{\partial \mathfrak{F}}{\partial \mu_j} = \frac{\mathrm{Tr}\,(N_j \, e^{-\beta \mathfrak{H}})}{\mathrm{Tr}\,\{e^{-\beta \mathfrak{H}}\}}. \qquad (6.71)$$

Then, taking the derivative of (6.71) with respect to $\mu_{j'}$, we obtain

$$
-\frac{\partial^2 \mathfrak{F}}{\partial \mu_j \partial \mu_{j'}} = \frac{1}{(\mathrm{Tr}\,\{e^{-\beta\mathfrak{H}}\})^2} \left[\beta\,\mathrm{Tr}\left\{N_j\,N_{j'}\,e^{-\beta\mathfrak{H}}\right\} \cdot \mathrm{Tr}\left\{e^{-\beta\mathfrak{H}}\right\} \right.
$$
$$
\left. - \mathrm{Tr}\{N_j e^{-\beta\mathfrak{H}}\} \cdot \mathrm{Tr}\{N_{j'} e^{-\beta\mathfrak{H}}\} \right]
$$
$$
= \beta[\langle N_j N_{j'}\rangle - \langle N_j\rangle\langle N_{j'}\rangle] = \beta[\langle (N_j - \langle N_j\rangle)(N_{j'} - \langle N_{j'}\rangle)\rangle]. \quad (6.72)
$$

Therefore, if we use the notation

$$
\mathfrak{N} := \sum_{j=1}^{s} (N_j - \langle N_j\rangle)\,\Delta\mu_j , \quad (6.73)
$$

we have that

$$
\sum_{j,j'} \left[-\frac{\partial^2 \mathfrak{N}}{\partial \mu_j \partial \mu_{j'}} \right] \Delta\mu_j\,\Delta\mu_{j'} = \beta\langle \mathfrak{N}^2\rangle . \quad (6.74)
$$

The next goal is to show that

$$
\langle \mathfrak{N}^2\rangle = 0 , \quad (6.75)
$$

only if all $\Delta\mu_j,\ j = 1, 2, \ldots, s$ vanish. Since the Hamiltonian H and the operators $\mathfrak{H}$ and N_j commute, the vectors ϕ_ω are also their eigenvectors, meaning that

$$
\mathfrak{H}\phi_\omega = \varepsilon_\omega \phi_\omega , \qquad \mathfrak{N}\phi_\omega = \nu_\omega \phi_\omega . \quad (6.76)
$$

Then, definitions (6.57) and (6.73) along (6.76) give

$$
\varepsilon_\omega = E_\omega - \sum_{j=1}^{s} \mu_j\, n_j^\omega , \quad (6.77)
$$

$$
\mathfrak{N}_\omega = \sum_{j=1}^{s} (n_j^\omega - \langle N_j\rangle)\Delta\mu_j . \quad (6.78)
$$

We can readily show that the mean value of the operator $\langle \mathfrak{N}^2\rangle$ on the Gibbs state $\mathfrak{F}$ is given by

$$
\langle \mathfrak{N}^2\rangle = \sum_\omega \left[\frac{e^{-\beta\varepsilon_\omega}}{\sum_{\omega'} e^{-\beta\varepsilon_{\omega'}}} \right] \cdot \nu_\omega^2 . \quad (6.79)
$$

This means that (6.75) holds if and only if $\nu_\omega = 0$ for all labels ω. This happens if and only if for all possible values of ω, we have that

$$\sum_{j=1}^{s} (n_j^\omega - \langle N_j \rangle)\Delta\mu_j = 0, \qquad (6.80)$$

and this latter condition is fulfilled if $\Delta\mu_j = 0$ for all values of j. This proves the uniqueness of the values of μ_j at equilibrium.

6.4.2 Thermodynamic Functions

The mean value of the operator $\mathfrak{H}$ in the Gibbs state is given by the expression

$$\frac{\partial}{\partial\beta}(\beta\mathfrak{F}) = \frac{\operatorname{Tr}[\mathfrak{H}\,e^{-\beta\mathfrak{H}}]}{\operatorname{Tr}[e^{-\beta\mathfrak{H}}]} = \langle\mathfrak{H}\rangle. \qquad (6.81)$$

This equation trivially gives

$$\langle\mathfrak{H}\rangle = \mathfrak{F} + \beta\,\frac{\partial\mathfrak{F}}{\partial\beta} \qquad (6.82)$$

or

$$\langle H \rangle = \mathfrak{F} - \beta\,\frac{\partial\mathfrak{F}}{\partial\beta} + \sum_{j=1}^{s}\mu_j n_j. \qquad (6.83)$$

In addition, there exists a relation between free energies for the canonical and grand canonical ensembles given by

$$F = \mathfrak{F} - \sum_{j=1}^{s}\mu_j\,\frac{\partial\mathfrak{F}}{\partial\mu_j}, \qquad (6.84)$$

where we have used (6.61).

6.4.3 Some Remarks on the Thermodynamic Definitions

The object of our study has been a Hamiltonian system, with s different species of identical particles. In the canonical representation, and for a classical system, we have

- The Gibbs state is $\mathcal{D} = \frac{1}{N_0}\,e^{-\beta H(\Omega)}$.

- The phase space Ω depends on the positions and momenta of all particles in the system. Thus, each point in the phase space is written as

$$\Omega = \Omega(\ldots, q_{j,a}, \ldots; \ldots p_{j,a}, \ldots),$$

where the index a labels the different species of particles and runs out from 1 to s. If the species labeled by the index a has N_a particles, then the index j varies between 1 and N_a.
- The free energy is given by

$$F = -\frac{1}{\beta} \ln \left\{ \frac{1}{n_1! \ldots n_s!} \frac{1}{h_0^{3N}} \int e^{-\beta H(\Omega)} \, d\Omega \right\},$$

where h_0 is the unit volume in phase space and

$$d\Omega = \prod_{a=1}^{s} \prod_{j=1}^{N_a} dq_{a_j} \, dp_{a_j}$$

is the Lebesgue measure in an Euclidean space of $6^{n_1 + \cdots + n_s}$.

Analogously, for the grand canonical ensemble the free energy is given by

$$\mathfrak{F} = -\frac{1}{\beta} \ln \left\{ \sum_{n_1 \ldots n_s} \left(\frac{1}{n_1! \ldots n_s!} \right) e^{-\beta \sum_j \mu_j n_j} \frac{1}{h_0^{3N}} \int d\Omega_{N_1} \ldots d\Omega_{N_s} \, e^{-\beta H(\Omega_{N_1}, \ldots, \Omega_{N_s})} \right\}.$$

The Statistical Mechanics of Unstable Quantum States

7

7.1 Introduction

We devote this chapter to an issue that has rarely been treated in the Literature: the possibility of assigning thermodynamic variables to unstable quantum systems. The interest of this study lies in the fact that many of existing quantum states are unstable, i.e., resonances in atomic, nuclear or particle physics [23].

In general, texts on quantum mechanics or quantum statistical mechanics deal with the statistical properties of bound states. As we have seen in previous chapters, time evolution of density operators constructed in the basis of Hermitian Hamiltonians is given by a Liouville-von Neumann equation, $i \dot{\rho} = [H, \rho]$ (we have set $\hbar = 1$ in all equations of this chapter).

Quantum unstable states are states which decay exponentially with time. Examples of these states are the solutions of scattering processes where one find states with complex values of the wave number, i.e., states with complex energies [23].

Actually in realistic unstable quantum systems the exponential decay is observed for most times. However, there are deviations of the exponential mode for very short (Zeno effect) as well as for very large (Khalfin effect) values of time. Nevertheless, these deviations are very hard to be observed and there are very few reports on these observations. Consequently, the exponential decay may be considered as a good approximation for most observations.

Then, one needs a mathematical object for the description of quantum unstable states in the same way that one uses vectors in Hilbert spaces or density matrices to describe stable states. In addition, we have to define a non-unitary time evolution to show the exponential decay of these states.

Assume that $t = 0$ is a time at which the decaying process starts. For simplicity, assume that the initial state is given by a pure state represented by the vector $|\psi\rangle$. At some later time, the vector state has evolved to $|\psi(t)\rangle = e^{-itH} |\psi\rangle$, where H is the Hamiltonian that governs the time evolution. The transition amplitude at time $t \geq 0$

O. Civitarese and M. Gadella, *Methods in Statistical Mechanics*, Lecture Notes in Physics 974, https://doi.org/10.1007/978-3-030-53658-9_7

is given by

$$A(t) := \langle \psi | e^{-itH} | \psi \rangle \,, \tag{7.1}$$

so that the transition probability at time t is given by

$$P(t) = |\langle \psi | e^{-itH} | \psi \rangle|^2 \,. \tag{7.2}$$

Note that $|\psi\rangle$ must be a scattering state for the Hamiltonian H, otherwise $P(t)$ would have been either a constant or an oscillatory function of time, for which

$$\lim_{t \to +\infty} A(t) = 0 \quad \Longrightarrow \quad \lim_{t \to +\infty} P(t) = 0 \,. \tag{7.3}$$

It is generally accepted that if $P(t)$ has the exponential form $e^{-t\Gamma}$ with $\Gamma > 0$, then $|\psi\rangle$ should represent a decaying state. In fact, there are vectors having this approximate behavior. Deviations from the exponential mode for normalized vectors, for very short and very large values of time (the abovementioned Zeno and Khalfin effects) follow directly from these considerations. For all other times, exponential decay dominates except for minor deviations due to the influence of environment (such as decoherence, multiple scattering, etc).

We have already noted that the deviation for short and large times is difficult to be observed, so that the pure exponential decay is an excellent approximation for most purposes. It has been proven that the vector state $|\psi\rangle$ can be split up as a sum of two contributions as $|\psi\rangle = |\psi^D\rangle + |\psi^B\rangle$, where $|\psi^D\rangle$ decays exponentially at all times $t \geq 0$ and $|\psi^B\rangle$ represents some "background" states that are supposed to be responsible for the deviations from the pure exponential law. It seems reasonable to expect that the time evolution of $|\psi^D\rangle$ be governed by the Hamiltonian H. From a conservative and practical point of view, we may desire that the time evolution for $|\psi^D\rangle$ be produced by an operator similar to e^{-itH}. However, if $e^{-itH}|\psi^D\rangle$ has to decay, the operator e^{-itH} has to loose its unitarity in some sense.

In 1958, N. Nakanishi proposed that $|\psi^D\rangle$ should be an eigenvector of the Hamiltonian H with complex eigenvalue so that $H|\psi^D\rangle = (E_R - i\Gamma/2)|\psi^D\rangle$, where E_R is the resonance energy (difference between the energy of the undecayed system and its decay products) and Γ is the inverse of the mean life [25]. Then, formally, for any $t \geq 0$ one has that

$$e^{-itH}|\psi^D\rangle = e^{-itE_R} e^{-t\Gamma} |\psi^D\rangle \,. \tag{7.4}$$

Taking the expression $H|\psi^D\rangle = (E_R - i\Gamma/2)|\psi^D\rangle$ as a definition for the state $|\psi^D\rangle$, we note that it fulfills the requirement of pure exponential decay for positive values of time. Now, the evolution operator e^{-itH} makes sense for $t \geq 0$ only and due to the decay mode given by (7.4), there is no sense to speak about the unitarity of e^{-itH} in this case.

Definition.- We call the vector $|\psi^D\rangle$ a *decaying Gamow vector*

This description has the advantage of describing the purely exponential decay part of a resonance in the familiar form of a vector state. However, some clear inconsistencies appear:

1. A Hamiltonian is a self-adjoint operator on a given Hilbert space and self-adjoint operators do not have complex eigenvalues. Therefore, $|\psi^D\rangle$ cannot be a regular normalizable vector in this Hilbert space.

2. On $|\psi^D\rangle$, the evolution operator e^{-itH} looses its unitarity and its group structure, since it makes sense for $t \geq 0$ only. This is a semigroup property. In addition, the mathematical meaning of (7.4) is not well defined.

All these inconsistencies have a cure, which is the use of Rigged Hilbert Spaces (RHS), also called Gelfand triplets [24]. Let us first define the concept of RHS and then discuss some consequences

(a) An RHS is a term of spaces

$$\Phi \subset \mathcal{H} \subset \Phi^\times , \tag{7.5}$$

where:

(i) The Hilbert space $\mathcal{H}$ is of infinite dimension with a countable orthonormal basis.

(ii) Φ is a dense subspace of $\mathcal{H}$. By dense, we mean that for any $\varphi \in \mathcal{H}$, there exists a sequence $\{\varphi_j\}$ of vectors in Φ that converges to φ in the usual sense of the convergence in Hilbert spaces. In addition, Φ has its own topology, compatible with the structure of linear space, such that the canonical mapping $i : \Phi \longmapsto \mathcal{H}$, defined for all $\varphi \in \Phi$ as $i(\varphi) = \varphi$, is continuous.

(iii) An antilinear functional on Φ is a mapping $F : \Phi \longmapsto \mathbb{C}$, where $\mathbb{C}$ is the field of complex numbers, such that for any φ, ψ in Φ and for any pair of complex numbers α and β, one has

$$F(\alpha \varphi + \beta \psi) = \alpha^* F(\varphi) + \beta^* F(\psi) , \tag{7.6}$$

where the star denotes complex conjugation. In addition, an antilinear functional is continuous if it is continuous as a mapping between the topological spaces Φ and $\mathbb{C}$. Then, $\Phi^\times$ is the set of all continuous antilinear functionals on Φ. This set has the form of a vector space, since linear combinations of the type $\alpha F + \beta G$, where $F, G \in \Phi^\times$ and $\alpha, \beta \in \mathbb{C}$ are also in $\Phi^\times$. The action of $\alpha F + \beta G$ on $\varphi \in \Phi$ is given by $\alpha F(\varphi) + \beta G(\varphi)$. We say that $\Phi^\times$ is the dual space of Φ. Incidentally, we may endow $\Phi^\times$ with a topology compatible with its character of dual space of Φ.

(b) RHS was used to implement the Dirac formulation of quantum mechanics. In particular, it includes plane waves and eigenvectors of observables for eigenvalues in the continuous spectrum as legitimate objects. Here, we show how we may use them to build a mathematically consistent picture of the resonance vector states and their time evolution.

7.1.1 Definition of Quantum Resonances in Non-relativistic Quantum Mechanics

The simplest and most popular picture of resonance phenomena relays in the idea according to which resonances are produced by an interaction on an otherwise stable state. This interaction would be responsible for the decay. Generally speaking, such an interaction is produced by forces derived from some potential V. Therefore, we need a Hamiltonian pair $\{H_0, H\}$, where $H = H_0 + V$.

In a scattering process, the incoming state interacts with the potential producing a quasi-stable state. Then, this quasi-stable state decays, originating the resonance.

Resonances are identified in scattering processes by the following features:

1. Bumps in the cross section centered at the resonance energy E_R with a width $\Gamma/2$. The inverse of the width accounts for the mean life.

2. Abrupt change of the phase shift at the value E_R.

3. The energy distribution of the decaying state obeys the Breit-Wigner form, also noticed by Fock, which is given by

$$f(E)\, dE = \frac{\Gamma^2/4}{(E - E_R)^2 + \Gamma^2/4}\, dE\,. \tag{7.7}$$

In fact, a necessary and sufficient condition for a state to have a purely exponential decay is that its energy distribution has exactly the form (7.7). In practice this is impossible for semi-bounded Hamiltonians, i.e., their spectrum has a finite lower bound and this has something to do with the stability of matter. This is another argument against the consideration of Gamow states as vectors in a physical Hilbert space.

Details on this definitions are found in the literature on scattering theory.

The mathematical characterizations of resonances are essentially two. Let us make a brief description of both, which suits for our purposes. The first one comes after the notion of resolvent operator.

1. Let H be a self-adjoint operator on a Hilbert space $\mathcal{H}$ and z a complex number not in the spectrum of H. Then, the operator $(H - zI)^{-1}$, where I is the identity on $\mathcal{H}$, is well defined (and bounded) and is called the *resolvent operator of H*. Usually one omits I and simply writes $(H - z)^{-1}$. For any vector $|\psi\rangle \in \mathcal{H}$, the function $\langle\psi|(H - z)^{-1}|\psi\rangle$ is an analytic function of one complex variable.

Let us also assume for simplicity that H has a pure (absolutely) continuous spectrum. This spectrum is the positive semi-axis, $\mathbb{R}^+ \equiv [0, \infty)$. The analytic function $\langle\psi|(H - z)^{-1}|\psi\rangle$ admits an analytic continuation through $[0, \infty)$.

Now, let us consider a Hamiltonian pair $\{H_0, H\}$ and assume that there exists a vector $|\psi\rangle \in \mathcal{H}$ such that $\langle\psi|(H_0 - z)^{-1}|\psi\rangle$ admits analytic continuation through $[0, \infty)$ with no singularities, and then $\langle\psi|(H - z)^{-1}|\psi\rangle$ also admits analytic continuation with poles (z_R). These poles represent resonances, being their real part of each pole the resonance energy and being their imaginary part the inverse of their mean lives. In general, these poles come into complex conjugate pairs (z_R, z_R^*).

2. In a scattering experiment there is a relation between the free asymptotic incoming and outgoing states given by the scattering matrix (or scattering operator). If the free incoming state is given in the momentum representation by $\psi^{in}(k)$ and the free outgoing state is $\psi^{out}(k)$, then this relation is often written as

$$\psi^{out}(k) = S(k)\,\psi^{in}(k)\,. \tag{7.8}$$

As a function of the momentum k, the function $S(k)$, also known as the scattering matrix or the S-operator, admits analytic continuation to the whole complex plane. This continuation has singularities in the form of poles (eventually branch cuts but never essential singularities). Pairs of poles symmetrically located with respect to the imaginary axis in the lower half plane give the resonances of the scattering process in terms of momenta.

If we go to the energy representation using the formula $E = \hbar^2 k^2/(2m)$, each pair of these poles is represented by two complex conjugate values of the form $E_R \pm i\Gamma/2$, where E_R is the resonance energy and $2/\Gamma$ the mean life.

These two definitions of resonance are not strictly equivalent, although their equivalence can be demonstrated for some models like the Friedrichs model, which be introduced later in this chapter.

7.1.2 More on Gamow Vectors

Here, we give an idea on how we may define Gamow vectors. First of all, let us use a notation which has been taken from the Dirac bra-ket notation. Let us go back to the concept of rigged Hilbert space (7.5). We represent the action of any functional $F \in \Phi^\times$ on a vector $\varphi \in \Phi$ either as $F(\varphi)$ or $\langle\varphi|F\rangle$. The latter generalizes the Dirac notation, because for any vector $\psi \in \mathcal{H}$, there are always a unique functional F_ψ such that $\langle\varphi|F_\psi\rangle = \langle\varphi|\psi\rangle$, where the last bracket denotes the Hilbert space scalar product.

Now, let us assume that H is a Hermitian operator with the property that for any vector φ in Φ, we have that $H\varphi$ is also in Φ, i.e., $H\Phi \subset \Phi$. We say that H leaves Φ invariant or that Φ reduces H. In addition, the operator H is continuous on Φ. Then, H can be extended into $\Phi^\times$ by means of the so-called duality formula, valid for all $\varphi \in \Phi$ and all $F \in \Phi^\times$:

$$\langle H\varphi|F\rangle = \langle\varphi|HF\rangle\,. \tag{7.9}$$

This defines the functional HF for all $F \in \Phi^\times$. The condition on the continuity of H on Φ is required because we want HF to be a continuous mapping from Φ into the set $\mathbb{C}$ of complex numbers. In this presentation, we omit any discussion concerning topologies. The conclusion is that we have extended H to be a linear mapping on $\Phi^\times$.

If H were not Hermitian, but its formal adjoint $H^\dagger$ leaves Φ invariant, i.e., $H^\dagger\Phi \subset \Phi$ and, furthermore, $H^\dagger$ is continuous on Φ, then we may extend H to $\Phi^\times$ by using a similar duality formula, valid for any $\varphi \in \Phi$ and any $F \in \Phi^\times$:

$$\langle H^\dagger \varphi | F \rangle = \langle \varphi | H F \rangle \,. \tag{7.10}$$

In the following, we shall use both (7.9) and (7.10).

We have already mentioned that resonance poles appear in pairs. Each pair represents the same resonance. In the energy representation, the pairs are complex conjugate of each other, so that they have the form $E_R \pm i\Gamma/2$, with $E_R, \Gamma > 0$. Let us use the notation $z_R = E_R - i\Gamma/2$, so that $z_R^* = E_R + i\Gamma/2$.

Now, let $H = H_0 + V$ be the Hamiltonian having the interaction that produces the resonance with resonance poles at z_R and z_R^*. With the use of Schwartz functions, Hardy functions on the positive (+) or the negative (−) half planes and the spectral theorem for self-adjoint operators, we construct two rigged Hilbert spaces

$$\Phi_\pm \subset \mathcal{H} \subset \Phi_\pm^\times \,, \tag{7.11}$$

with the following properties:

1. For any $\varphi_\pm \in \Phi_\pm$, we have that $H\varphi_\pm \in \Phi_\pm$. In addition, H is continuous on $\Phi_\pm$, so that H can be extended to the respective dual $\Phi_\pm^\times$ using the duality formula (7.9).

2. There exists a vector $|\psi^D\rangle \in \Phi_+^\times$ and another vector $|\psi^G\rangle \in \Phi_-^\times$ such that

$$H |\psi^D\rangle = z_R |\psi^D\rangle \,, \qquad H |\psi^G\rangle = z_R^* |\psi^G\rangle \,. \tag{7.12}$$

The meaning of $|\psi^G\rangle$ will be clarified soon. The first eigenvalue equation in (7.12) is valid in $\Phi_+^\times$ and the second in $\Phi_-^\times$.

3. The evolution semigroup e^{itH} leaves Φ_+ invariant for values of time $t \geq 0$ and leaves Φ_- invariant for $t \leq 0$, so that

$$e^{itH} \Phi_+ \subset \Phi_+ \,, \quad t \geq 0; \qquad e^{itH} \Phi_- \subset \Phi_- \,, \quad t \leq 0. \tag{7.13}$$

Taking into account the duality formula (7.10) and that the formal adjoint of e^{itH} is e^{-itH}, the evolution operator e^{-itH} is extended to the duals:

$$\langle e^{itH} \varphi_\pm | F_\pm \rangle = \langle \varphi_\pm | e^{-itH} F_\pm \rangle \,, \tag{7.14}$$

with $\varphi_\pm \in \Phi_\pm$, $F_\pm \in \Phi_\pm^\times$ and (7.14) is valid for $t \geq 0$ with plus sign and for $t \leq 0$ with minus sign.

Once we have extended the time evolution to the duals $\Phi_\pm^\times$, we can determine the time evolution for $|\psi^D\rangle$ and $|\psi^G\rangle$, which is given by

$$e^{-itH} |\psi^D\rangle = e^{-itE_R} e^{-t\Gamma/2} |\psi^D\rangle \,, \quad \text{if} \quad t \geq 0, \tag{7.15}$$

$$e^{-itH} |\psi^G\rangle = e^{-itE_R} e^{t\Gamma/2} |\psi^G\rangle \,, \quad \text{if} \quad t \leq 0. \tag{7.16}$$

Then, the exponential decay has been shown for $|\psi^D\rangle$ for all positive values of time. On the other hand, the vector $|\psi^G\rangle$ grows exponentially as $t \longmapsto 0$, but decays exponentially as $t \longmapsto -\infty$. We have already remarked that the resonance poles z_R

and its complex conjugate z_R^* represent the same resonance, since what is meaningful in a resonance are E_R and Γ. Then, in principle, the vectors $|\psi^D\rangle$ and $|\psi^G\rangle$ should be equally capable to exhibit the exponentially decaying part of a resonance. We solve this question below.

The Gamow vector $|\psi^D\rangle$ will be called the *decaying Gamow vector*, while $|\psi^G\rangle$ will be the *growing Gamow vector*. These names hide a kind of prejudice: that the time flows in one direction only from $-\infty$ to ∞. Along this direction, $|\psi^G\rangle$ grows with time and $|\psi^D\rangle$ decays with time, and this interpretation has been advocated by Prigogine and the Brussels school.

4. With respect to the time reversal operator T, all these objects have the following behavior:

$$T\Phi_\pm = \Phi_\mp, \quad T\Phi_\pm^\times = \Phi_\mp^\times, \quad T|\psi^D\rangle = |\psi^G\rangle, \quad T|\psi^G\rangle = |\psi^D\rangle, \quad (7.17)$$

and also

$$T e^{-itH}|\psi^D\rangle = e^{-i(-t)H}|\psi^G\rangle, \quad t \geq 0; \qquad T e^{-itH}|\psi^G\rangle = e^{-i(-t)H}|\psi^D\rangle, \quad t \leq 0. \quad (7.18)$$

This shows that the growing Gamow vector represents exactly the same object than the decaying Gamow vector in a picture in which the time flow has been reversed. In fact, processes with plus sign evolve forward with time and processes with sign minus evolve backward with time.

Nevertheless, another interpretation is possible, which is given by the so-called Time Asymmetric Quantum Mechanics (TAQM), introduced by A. Bohm as an attempt to define a quantum arrow of time. In TAQM, vectors in the triplet $\Phi_- \subset \mathcal{H} \subset \Phi_-^\times$, known as *in*-vectors, are produced in the past by a *preparation apparatus*, before the scattering process that produces the resonance has taken place. These vectors represent *states* and make sense for $t < 0$ only.

After the production and the decay of the resonance, the decaying products go far apart of the interaction region and are observed. Then, vectors of the triplet $\Phi_+ \subset \mathcal{H} \subset \Phi_+^\times$, called *out*-vectors, are interpreted as observables which are measured when the interaction has ceased by a *registration apparatus* and considered for $t > 0$ only. In this TAQM interpretation, the arrow of times goes from the preparation to the registration apparatus.

7.2 The Friedrichs Model

There is a model for resonances that can be presented in simple, yet fully rigorous mathematical terms with the additional and important advantage of being exactly solvable. This is the Friedrichs model that we briefly describe in the sequel. In the basic version of the Friedrichs model [25], a bound state interacts with an external field in such a way that the bound state becomes unstable, i.e., it becomes a resonance.

The spectrum of the unperturbed Hamiltonian H_0 has a non-degenerate bound state of energy $\omega_0 > 0$ and a continuous spectrum and it admits a spectral decomposition of the form

$$H_0 = \omega_0 |1\rangle\langle 1| + \int_0^\infty \omega |\omega\rangle\langle\omega| \, d\omega \,, \tag{7.19}$$

where $H_0 |1\rangle = \omega_0 |1\rangle$. The vectors $|\omega\rangle$ are eigenvectors of H_0 with eigenvalue ω in the continuous spectrum of H_0, so that $H_0 |\omega\rangle = \omega |\omega\rangle$. These eigenvectors are not normalizable vectors in Hilbert space but instead functionals on a suitable rigged Hilbert space, as explained in the previous section. The set of vectors $\{|1\rangle, |\omega\rangle)\}$, with $\omega \in \mathbb{R}^+ \equiv [0, \infty)$ serves as a basis for the Hilbert space $\mathcal{H}$ on which H_0 acts, so that any vector $\varphi \in \mathcal{H}$ may be written as

$$\varphi = \alpha |1\rangle + \int_0^\infty h(\omega) |\omega\rangle \, d\omega \,, \quad \text{with} \quad \alpha \in \mathbb{C}, \quad \int_0^\infty |h(\omega)|^2 \, d\omega < \infty \,. \tag{7.20}$$

Being given two vectors $\varphi, \phi \in \mathcal{H}$ with φ as in (7.20) and $\phi = \beta |1\rangle + \int_0^\infty g(\omega) |\omega\rangle \, d\omega$, their scalar product is given by

$$\langle\varphi|\phi\rangle = \alpha^*\beta + \int_0^\infty h^*(\omega) \, g(\omega) \, d\omega \,, \tag{7.21}$$

which can be formally obtained if we use the following rules:

$$\langle 1|1\rangle = 1 \,, \qquad \langle 1|\omega\rangle = \langle\omega|1\rangle = 0 \,, \qquad \langle\omega|\omega'\rangle = \delta(\omega - \omega') \,. \tag{7.22}$$

The potential V intertwines discrete and continuous spectrum of H_0 with a weight function, $f(\omega)$, called the *form factor*. Usually, $f(\omega)$ is square integrable on $\mathbb{R}^+ = [0, \infty)$. Then for real $f(\omega)$, V, has the following form:

$$V = \int_0^\infty f(\omega) \left[|\omega\rangle\langle 1| + |1\rangle\langle\omega| \right] d\omega \,. \tag{7.23}$$

The interacting Hamiltonian $H = H_0 + \lambda V$ depends on the coupling constant λ and on the form factor $f(\omega)$. Under simple conditions on $f(\omega)$, H has the following properties:

1. The Hamiltonian H has purely (simple, absolutely) continuous spectrum, which is equal to $\mathbb{R}^+ \equiv [0, \infty)$.
2. The function $\eta(z)$ defined by

$$\frac{1}{\eta(z)} := \langle 1|\frac{1}{z - H}|1\rangle \tag{7.24}$$

is analytic on the complex plane with a branch cut on the positive semi-axis $\mathbb{R}^+$ and no zeroes. It admits an analytic continuation through the cut from above to below, on the lower half plane, which has a unique simple zero located at a point $z_R = E_R - i\Gamma/2$, with E_R, $\Gamma > 0$, which we may determine explicitly. Analogously, $\eta(z)$ has another analytic continuation through the cut from below to above, on the upper half plane, with a unique simple zero at $z_R^* = E_R + i\Gamma/2$.

The expression (7.24) is called the *reduced resolvent* of H. It has no poles, but their analytic continuations through the cut have simple poles at the points z_R and z_R^*.

Then, the situation is the following: We have found a vector $|\psi\rangle \equiv |1\rangle$ for which $\langle\psi|(z - H)^{-1}|\psi\rangle$ admits analytic continuations through the cut with simple poles at z_R and z_R^*. On the contrary, $\langle\psi|(z - H_0)^{-1}|\psi\rangle = \langle 1|(z - H_0)^{-1}|1\rangle = (z - \omega_0)^{-1}$ does not have such poles. According to the definition given in Sect. 7.1.1, z_R and z_R^* are resonance poles.

Now, we want to give explicit expressions of z_R, z_R^* and their corresponding Gamow vectors. For a function $F(\omega)$, let us use the following notation:

$$\int_0^\infty d\omega \, \frac{\lambda^2 F^2(\omega)}{z_R - \omega - i0} := \int_0^\infty d\omega \, \frac{\lambda^2 F^2(\omega)}{z_R - \omega} - 2\pi i \lambda^2 |F(z_R)|^2 \qquad (7.25)$$

and

$$\int_0^\infty d\omega \, \frac{\lambda^2 F^2(\omega)}{z_R^* - \omega + i0} := \int_0^\infty d\omega \, \frac{\lambda^2 F^2(\omega)}{z_R^* - \omega} + 2\pi i \lambda^2 |F(z_R^*)|^2 . \qquad (7.26)$$

Then, one can prove that

$$z_R = \omega_0 + \int_0^\infty d\omega \, \frac{\lambda^2 f^2(\omega)}{z_R - \omega - i0} , \qquad z_R^* = \omega_0 + \int_0^\infty d\omega \, \frac{\lambda^2 f^2(\omega)}{z_R^* - \omega + i0} . \qquad (7.27)$$

Also, we may write the explicit form of the Gamow vectors as

$$|\psi^D\rangle = |1\rangle + \int_0^\infty d\omega \, \frac{\lambda f(\omega)}{z_R - \omega - i0} |\omega\rangle , \qquad (7.28)$$

$$|\psi^G\rangle = |1\rangle + \int_0^\infty d\omega \, \frac{\lambda f(\omega)}{z_R^* - \omega + i0} |\omega\rangle . \qquad (7.29)$$

These formulas for the decaying and growing Gamow vectors should be understood as functionals on the spaces Φ_+ and Φ_-, respectively. Then, for any $\varphi_+ \in \Phi_+$, we apply (7.28) to have the following expression:

$$\langle\varphi_+|\psi^D\rangle = \langle\varphi_+|1\rangle + \int_0^\infty d\omega \, \frac{\lambda f(\omega)}{z_R - \omega - i0} \langle\varphi_+|\omega\rangle . \qquad (7.30)$$

All ingredients in (7.30) are well defined. Since $\Phi_+ \subset \mathcal{H}$, the term $\langle\varphi_+|1\rangle$ is just the scalar product of these vectors on $\mathcal{H}$. For $\omega \in \mathbb{R}^+$, it is shown that $\langle\varphi_+|\omega\rangle$ is a square-integrable function on $\mathbb{R}^+$. Analogously, a similar expression is valid for $|\psi^G\rangle$ acting on Φ_-.

For each ω in the continuous spectrum, we may find a pair of functionals $|\omega_\pm\rangle \in \Phi_\pm^\times$ such that

$$H\,|\omega_\pm\rangle = \omega\,|\omega_\pm\rangle. \tag{7.31}$$

Now, recalling that $F_\pm \equiv |F_\pm\rangle \in \Phi_\pm^\times$ are continuous *antilinear* functionals on $\Phi_\pm$, we have that

$$\langle F_\pm | \varphi_\pm \rangle := \langle \varphi_\pm | F_\pm \rangle^*, \tag{7.32}$$

where the star denotes complex conjugation. Equation (7.32) defines $\langle F_\pm |$ as a continuous *linear* functional on $\Phi_\pm$.

With respect to the pair of rigged Hilbert spaces, $\Phi_\pm \subset \mathcal{H} \subset \Phi_\pm^\times$, the Hamiltonian H admits two spectral decompositions, which are written as

$$H = z_R \, |\psi^D\rangle\langle\psi^G| + \int_0^\infty \omega\,|\omega_+\rangle\langle\omega_-|\,d\omega, \tag{7.33}$$

and

$$H = z_R^* \, |\psi^G\rangle\langle\psi^D| + \int_0^\infty \omega\,|\omega_-\rangle\langle\omega_+|\,d\omega. \tag{7.34}$$

While H in (7.33) is a continuous linear mapping from Φ_- into $\Phi_+^\times$, H in (7.34) is a continuous linear mapping from Φ_+ into $\Phi_-^\times$. Furthermore, each spectral decomposition is the formal adjoint of the other. Note that, since $\Phi_\pm \subset \mathcal{H} \subset \Phi_\pm^\times$, we have that $\Phi_+ \subset \Phi_-^\times$ and $\Phi_- \subset \Phi_+^\times$.

Using (7.32), we can show that

$$\langle\psi^D|\,H = z_R^* \,\langle\psi_-^D|, \qquad \langle\psi^G|\,H = z_R \,\langle\psi^G|. \tag{7.35}$$

The proof of (7.35) is very simple. The action of the functional $H|\psi^D\rangle$ on the vector $\varphi_+ \in \Phi_+$ is $\langle\varphi_+|H|\psi^D\rangle = z_R\,\langle\varphi_+|\psi^D\rangle$. Taking the complex conjugate of this and using (7.32), we have that

$$\langle\psi^D|H|\varphi_+\rangle = z_R^*\,\langle\psi^D|\varphi_+\rangle. \tag{7.36}$$

Finally, it is also interesting to remark that there are explicit formulas for the eigenvectors $|\omega^\pm\rangle$ of the Hamiltonian H with eigenvalues $\omega \in \mathbb{R}^+$. These are

$$|\omega_\pm\rangle = |\omega\rangle + \frac{\lambda f(\omega)}{\eta_\pm(\omega)}\left(|1\rangle + \int_0^\infty d\omega' \,\frac{\lambda f(\omega')}{\omega - \omega' \pm i0}\,|\omega'\rangle\right), \tag{7.37}$$

where $\eta_\pm(\omega)$ will be defined in (7.46).

This completes a brief description of the Friedrichs model including the form of the Gamow vectors in the context of this model.

7.2.1   The Friedrichs Model in the Second Quantization Language

The Friedrichs model admits an alternative description in terms of creation and annihilation operators. One starts with a vacuum $|0\rangle$ and the creation operators $a^\dagger$ and $b_\omega^\dagger$, $\omega \in \mathbb{R}^+$, transforming $|0\rangle$ into $|1\rangle$ and $|\omega\rangle$, respectively. This is .

$$|1\rangle = a^\dagger |0\rangle\,, \qquad |\omega\rangle = b_\omega^\dagger |\omega\rangle\,. \tag{7.38}$$

The corresponding annihilation operators a and b_ω must satisfy the following commutation relations with the creation operators:

$$[a, a^\dagger] = 1\,, \qquad [b_\omega, b_\omega^\dagger] = \delta(\omega - \omega') \tag{7.39}$$

and satisfy the natural property

$$a\,|0\rangle = b_\omega\,|0\rangle = 0\,, \qquad \omega \in \mathbb{R}^+\,. \tag{7.40}$$

Then, in this language we write

$$H_0 = \omega_0\, a^\dagger\, a + \int_0^\infty d\omega\, \omega\, b_\omega^\dagger\, b_\omega\,, \tag{7.41}$$

$$V = \int_0^\infty d\omega\, f(\omega)(a^\dagger\, b_\omega + a\, b_\omega^\dagger)\,. \tag{7.42}$$

The interacting Hamiltonian $H = H_0 + V$ admits a decomposition in terms of creation and annihilation operators of Gamow states and some similar operators corresponding to the continuum. Now, assume that the form factor $f(\omega)$ may admit an analytic continuation to the lower half plane and defined around the point z_R. Then, let us define

$$A_{IN}^\dagger := \int_\gamma d\omega\, \frac{\lambda\, f(\omega)}{w - z_R}\, b_\omega^\dagger - a^\dagger\,, \tag{7.43}$$

$$A_{OUT} := \int_\gamma d\omega\, \frac{\lambda\, f(\omega)}{\omega - z_R}\, b_\omega - a\,, \tag{7.44}$$

where γ is a contour around z_R.

Then, go back to (7.24). The explicit form of the function $\eta(z)$ is

$$\eta(z) = \omega_0 - z - \int_0^\infty d\omega\, \frac{\lambda^2\, f^2(\omega)}{\omega - z}\,. \tag{7.45}$$

As was previously noted, this function is analytic on the complex plane and has a branch cut, which is the positive semi-axis $\mathbb{R}^+ \equiv [0, \infty)$. Then, for all real $\omega > 0$, set:

$$\eta_\pm(\omega) := \lim_{\epsilon \to 0} \eta(\omega \mp i\epsilon)\,. \tag{7.46}$$

These limits are different. Then, define the function $\widetilde{\eta}_+(\omega)$ as

$$\frac{1}{\widetilde{\eta}_+(\omega)} := \frac{1}{\eta_+(\omega)} + 2\pi i K \, \delta(w - z_R), \qquad (7.47)$$

where

(i) K is the residue of $1/\eta(z)$ at the pole z_R.

(ii) The expression $\delta(w - z_R)$ denotes a distribution on functions $f(\omega)$ defined on the positive semi-axis and having analytic continuation to the lower half plane, so that $\int_0^\infty f(\omega) \, \delta(\omega - z_R) \, d\omega = f(z_R)$.

Then for any $\omega \in \mathbb{R}^+$, we define the following operators:

$$B^\dagger_{\omega,IN} := b^\dagger_\omega + \frac{\lambda f(\omega)}{\widetilde{\eta}_+(\omega)} \left\{ \int_0^\infty d\omega' \, \frac{\lambda f(\omega')}{\omega' - \omega - i0} \, b^\dagger_{\omega'} - a^\dagger \right\}, \qquad (7.48)$$

$$B_{\omega,OUT} := b_\omega + \frac{\lambda f(\omega)}{\widetilde{\eta}_+(\omega)} \left\{ \int_0^\infty d\omega' \, \frac{\lambda f(\omega')}{\omega' - \omega - i0} \, b_{\omega'} - a \right\}. \qquad (7.49)$$

It is interesting to note that the above operators satisfy the following commutation relations:

$$[A_{OUT}, A^\dagger_{IN}] = 1, \qquad \frac{\eta_+(\omega)}{\eta_-(\omega)} [B_{\omega,OUT}, B^\dagger_{\omega',IN}] = \delta(\omega - \omega'), \qquad (7.50)$$

all other possible commutators vanish. Note that the operators A can be multiplied by each other and the same property is valid for the operators B. This is rather technical, but it comes from the fact that they all are defined in terms of distributional kernels on the lower half plane. These operators serve as creation and annihilation operators for the decaying Gamow vector:

$$|\psi^D\rangle = A^\dagger_{IN} |0\rangle, \qquad A_{OUT} |\psi^D\rangle = |0\rangle, \qquad (7.51)$$

and

$$|\psi^G\rangle = A^\dagger_{OUT} |0\rangle, \qquad A_{IN} |\psi^G\rangle = |0\rangle. \qquad (7.52)$$

Finally, the Hamiltonian H may be written in terms of the "A" as well as of the "B" operators in the following form:

$$H = z_R A^\dagger_{IN} A_{OUT} + \int_0^\infty d\omega \, \omega \, \frac{\eta_+(\omega)}{\eta_-(\omega)} B^\dagger_{\omega,IN} B_{\omega,OUT}, \qquad (7.53)$$

and

$$H = z_R^* A^\dagger_{OUT} A_{IN} + \int_0^\infty d\omega \, \omega \, \frac{\eta_-(\omega)}{\eta_+(\omega)} B^\dagger_{\omega,OUT} B_{\omega,IN}. \qquad (7.54)$$

These are second quantized forms of (7.33). We may proceed by analogy and establish a second quantized version of (7.34).

7.3   The Entropy for the Harmonic Oscillator from the Path Integral Approach

Our final objective is to determine the entropy for a quantum decaying system.

For any quantum system with Hamiltonian H and canonical partition function $Z := \operatorname{Tr} e^{-\beta H}$, $\beta := 1/(kT)$, where k is the Boltzmann constant and T the absolute temperature, the canonical entropy is given by the following expression:

$$S = k \left(1 - \beta \frac{\partial}{\partial \beta} \right) \log Z . \tag{7.55}$$

Then, assume that H is the Hamiltonian of the one-dimensional harmonic oscillator and call $|n\rangle$ to its eigenfunctions, $n = 0, 1, 2, \ldots$. The partition function for this system is now:

$$Z = \operatorname{Tr} e^{-\beta H} = \sum_{n=0}^{\infty} \langle n | e^{-\beta H} | n \rangle . \tag{7.56}$$

It is a simple exercise to find and explicit expression for the canonical entropy, using (7.55). This gives

$$S = -k \, \log[2 \, \sinh(\beta \hbar \omega / 2)] + k \, \frac{\beta \hbar \omega}{2} \, \coth \left(\frac{\beta \hbar \omega}{2} \right) . \tag{7.57}$$

We have adopted the path integrals to write the partition function in a basis of coherent states, as explained in previous chapters. To start with, we express the canonical ensemble in terms of coherent states, and then use path integrals to compute the density operator. As is well known, coherent states are defined from a vacuum state $|0\rangle$ as

$$|\alpha\rangle := e^{\alpha a^{\dagger}} |0\rangle . \tag{7.58}$$

In our case, $a^{\dagger}$ and a are the creation and annihilation operators for the harmonic oscillator, respectively, acting on the vacuum state $|0\rangle$. The states (7.58) are eigenstates of the annihilation operator a with eigenvalues α. Take now the density operator $\rho = e^{-\beta H}$ and use the strategy of path integrals to estimate its matrix elements with respect to the coherent states (7.58). Then, for any pair of complex numbers α_i and α_f, let us define

$$\langle \alpha_i | \rho | \alpha_f \rangle := \frac{1}{\langle \alpha_i | \alpha_f \rangle} \lim_{N \to \infty} \rho_N(\alpha_i, \alpha_f) , \tag{7.59}$$

where

$$\rho_N(\alpha_i, \alpha_f) = \int \prod_{k=1}^{N} \left(\frac{d^2 \alpha_k}{\pi} \right) \exp \left\{ -\tau \left[\sum_{n=1}^{N} H_+(\alpha_{n-1}, \alpha_n) \right. \right.$$
$$\left. \left. + \sum_{n=1}^{N+1} \left\{ \left(\frac{\alpha_n^* - \alpha_{n-1}^*}{2\tau} \right) \alpha_n - \alpha_{n-1}^* \left(\frac{\alpha_n - \alpha_{n-1}}{2\tau} \right) \right\} \right] \right\} . \tag{7.60}$$

Here, $\alpha_i = \alpha_0$, $\alpha_f = \alpha_{N+1}$ (the subindices i and f stand here for "initial" and "final", respectively) and $\tau = \beta/N$, $k = 1, 2, \ldots, N$. Each of the α_k is complex, so that $\alpha_k = x_k + iy_k$ and $d^2\alpha_k = dx_k\, dy_k$. Thus, we have $2N$ integration variables, $x_1, x_2, \ldots, x_N$ and $y_1, y_2, \ldots, y_N$ and all integration limits are $-\infty$ and $+\infty$. In (7.55), factors of the form $H_+(\alpha, \alpha')$ appear having the following explicit form:

$$H_+(\alpha, \alpha') = \frac{\langle \alpha | H | \alpha' \rangle}{\langle \alpha | \alpha' \rangle}, \tag{7.61}$$

where

$$\langle \alpha | \alpha' \rangle = \exp\left\{ -\frac{|\alpha|^2}{2} - \frac{|\alpha'|^2}{2} + \alpha^* \alpha' \right\}. \tag{7.62}$$

In the particular case of the one-dimensional harmonic oscillator, the explicit form of (7.61) can be easily obtained, just by using the form of H in terms of creation and annihilation operators and taking into account that $a|\alpha\rangle = \alpha\,|\alpha\rangle$ so that $\langle \alpha | a^\dagger = \alpha^* \langle \alpha |$. This gives

$$H_+(\alpha, \alpha') = \hbar\omega(\alpha^*\alpha + 1/2). \tag{7.63}$$

Then, let us take the exponential in (7.60) and use (7.63). After some calculations, we obtain a product of five exponentials in the following form:

$$\exp\left\{ -\tau \left[\sum_{n=1}^{N} H_+(\alpha_{n-1}, \alpha_n) + \sum_{n=1}^{N+1} \left\{ \left(\frac{\alpha_n^* - \alpha_{n-1}^*}{2\tau} \right) \alpha_n - \alpha_{n-1}^* \left(\frac{\alpha_n - \alpha_{n-1}}{2\tau} \right) \right\} \right] \right\}$$

$$= \exp\left\{ -\frac{1}{2} \tau N \hbar\omega \right\} \times \exp\left\{ -\frac{1}{2}(|\alpha_i|^2 + |\alpha_f|^2) \right\} \times \exp\left\{ -\sum_{k=1}^{N} |\alpha_k|^2 \right\}$$

$$\times \exp\left\{ (1 - \tau\hbar\omega) \sum_{k=1}^{N} \alpha_{k-1}^* \alpha_k \right\} \times \exp\left\{ \alpha_N^* \alpha_{N+1} \right\}. \tag{7.64}$$

Note that $\tau = \beta/N$, so that the first product in the second line of (7.64) takes the form $\exp\{-\beta\hbar\omega/2\}$. Also, use the notation $\sigma := 1 - \tau\hbar\omega$. Then, we proceed with the integration of (7.60) step by step, starting with the integration with respect to α_1, i.e., the variables x_1 and y_1, then α_2 and so on. Note that the two first terms after the equal sign in (7.64) are constant, so that the integration concerns the other three only. Then, integration on α_1 gives

$$\frac{1}{\pi} \int_{-\infty}^{\infty} dx_1 \int_{-\infty}^{\infty} dy_1 \, \exp\{-x_1^2 - x_2^2 + \sigma(x_1 + iy_1)\alpha_i^* + \sigma(x_1 - y_1)\alpha_2\}$$

$$\frac{1}{\pi} \int_{-\infty}^{\infty} dx_1 \int_{-\infty}^{\infty} dy_1 \, \exp\left\{ -\left[x_1 - \frac{\sigma}{2}(\alpha_i^* - \alpha_2) \right]^2 + \frac{\sigma^2}{4}(\alpha_i^* + \alpha_2)^2 \right.$$

$$\left. -\left[y_1 - i\frac{\sigma}{2}(\alpha_i^* - \alpha_2) \right]^2 - \frac{\sigma^2}{4}(\alpha_i^* - \alpha_2)^2 \right\}. \tag{7.65}$$

This integral does not converge in strict sense due to the presence of the terms $i\sigma y_1(\alpha_i^* - \alpha_2)$ and $\sigma x_1(\alpha_i^* + \alpha_2)$ in the exponential. The usual cure consists in a "regularization" based in the elimination of the terms under the integral sign responsible for the divergencies. This transforms the last integral in (7.65) into

$$\frac{1}{\pi} \int_{-\infty}^{\infty} dx_1 \int_{-\infty}^{\infty} dy_1 \exp\left\{-x_1^2 + \frac{\sigma^2}{4}(\alpha_i^* + \alpha_2)^2 - y_1^2 - \frac{\sigma^2}{4}(\alpha_i^* - \alpha_2)^2\right\}$$
$$= \frac{1}{\pi} \exp\{\sigma^2\alpha_i^*\alpha_2\} \int_{-\infty}^{\infty} dx_1\, e^{-x_1^2} \int_{-\infty}^{\infty} dy_1\, e^{-y_1^2} = \exp\{\sigma^2\alpha_i^*\alpha_2\}. \qquad (7.66)$$

Once we have performed this "regularized" integration over α_1, we proceed with a similar integration over α_2. The integral in α_2 has the following form:

$$\frac{1}{\pi} \int_{-\infty}^{\infty} dx_2 \int_{-\infty}^{\infty} dy_2 \exp\left\{-\left[x_2 - \frac{\sigma}{2}(\sigma\alpha_i^* - \alpha_3)\right]^2 + \frac{\sigma^2}{4}(\sigma\alpha_i^* + \alpha_3)^2\right.$$
$$\left. -\left[y_2 - i\frac{\sigma}{2}(\sigma\alpha_i^* - \alpha_3)\right]^2 - \frac{\sigma^2}{4}(\sigma\alpha_i^* - \alpha_3)^2\right\}. \qquad (7.67)$$

Again, this integral diverges and has to be "regularized", so that we eliminate the terms with exponents of the form $\sigma x_2(\sigma\alpha_i^* - \alpha_3)$ and $i\sigma y_2(\sigma\alpha_i^* - \alpha_3)$, so as to give the following result:

$$\frac{1}{\pi} \exp\{\sigma^3\alpha_i^*\alpha_3\} \int_{-\infty}^{\infty} dx_2\, e^{-x_2^2} \int_{-\infty}^{\infty} dy_2\, e^{-y_2^2} = \exp\{\sigma^3\alpha_i^*\alpha_3\}. \qquad (7.68)$$

Continuing with this procedure and after integrating up to the variable α_{N-1}, we arrive to the term

$$\exp\{\sigma^{N-1}\alpha_i^*\alpha_N\}, \qquad (7.69)$$

which must be included in the integration on the last variable α_N. This last integral takes the form

$$\frac{1}{\pi} \int_{-\infty}^{\infty} dx_N \int_{-\infty}^{\infty} dy_N \exp\left\{-\left[x_N - \frac{1}{2}(\sigma^N\alpha_i^* + \alpha_f)\right]^2 + \frac{1}{4}(\sigma^N\alpha_i^* + \alpha_f)^2\right.$$
$$\left. -\left[y_N - \frac{i}{2}(\sigma^N\alpha_i^* - \alpha_f)\right]^2 - \frac{1}{4}(\sigma^N\alpha_i^* - \alpha_f)^2\right\}. \qquad (7.70)$$

In the regularization process, we drop the terms $-\frac{1}{2}(\sigma^N\alpha_i^* + \alpha_f)$ and $-\frac{i}{2}(\sigma^N\alpha_i^* - \alpha_f)$ in (7.65) and then integrate so as to obtain the following value for the regularized integral (7.65):

$$\exp\{\sigma^N\alpha_i^*\alpha_f\}. \qquad (7.71)$$

This gives the final form of (7.60), which is obtained after multiplication of (7.71) by the two first constant terms after the equal sign in (7.64). This is

$$\exp\left\{-\frac{1}{2}\beta\hbar\omega\right\} \times \exp\left\{-\frac{1}{2}(|\alpha_i|^2 + |\alpha_f|^2)\right\} \times \exp\{\sigma^N \alpha_i^* \alpha_f\}. \tag{7.72}$$

Recall that $\sigma = 1 - \tau\hbar\omega$ with $\tau = \beta/N$. Since $\hbar$ is very small, we may use the approximation given by $\sigma^N \approx 1 - N\tau\hbar\omega = 1 - \beta\hbar\omega$. This shows that (7.71) can be written as

$$\exp\left\{-\frac{1}{2}\beta\hbar\omega\right\} \times \exp\left\{-\frac{1}{2}(|\alpha_i|^2 + |\alpha_f|^2)\right\} \times \exp\left\{(1 - \beta\hbar\omega)\alpha_i^* \alpha_f\right\}$$

$$= \exp\left\{-\frac{1}{2}\beta\hbar\omega\right\} \times \exp\left\{-\frac{1}{2}(|\alpha_i|^2 + |\alpha_f|^2 - 2\alpha_i^* \alpha_f)\right\} \times \exp\left\{-\beta\hbar\omega\,\alpha_i^* \alpha_f\right\}$$

$$= \langle\alpha_i|\alpha_f\rangle \exp\left\{-\frac{1}{2}\beta\hbar\omega\right\} \times \exp\{-\beta\hbar\omega\,\alpha_i^* \alpha_f\} = \rho_N(\alpha_i, \alpha_f). \tag{7.73}$$

Observe that the expression in (7.73) does not depend on N. Then, we arrive to the final expression for (7.59), which is

$$\rho(\alpha_i, \alpha_f) := \langle\alpha_i|\rho|\alpha_f\rangle = \frac{1}{\langle\alpha_i|\alpha_f\rangle} \rho_N(\alpha_i, \alpha_f) = \exp\left\{-\frac{1}{2}\beta\hbar\omega\right\} \times \exp\{-\beta\hbar\omega\,\alpha_i^* \alpha_f\}. \tag{7.74}$$

Using coherent states, the partition function $Z \equiv \text{Tr}\,\rho$ may still be calculated using the following formula ($\alpha = x + iy$):

$$Z = \text{Tr}\,e^{-\beta H} = \int_{\mathbb{C}} \frac{d^2\alpha}{\pi} \langle\alpha|e^{-\beta H}|\alpha\rangle$$

$$= \int \frac{d^2\alpha}{\pi} \rho(\alpha, \alpha) = \frac{1}{\pi} e^{-(\beta\hbar\omega)/2} \int_{-\infty}^{\infty} dx \int_{-\infty}^{\infty} dy\, e^{-\beta\hbar\omega(x^2 + y^2)} = e^{-(\beta\hbar\omega)/2} \frac{1}{\beta\hbar\omega}, \tag{7.75}$$

where the integration over α is extended on the whole complex plane.

Now, use (7.55) to obtain the following result:

$$S = k(1 - \log(\beta\hbar\omega)). \tag{7.76}$$

This result has been the consequence of regularization and a simplification such as $(1 - \tau\hbar\omega)^N \approx 1 - N\tau\hbar\omega = 1 - \beta\hbar\omega$. Since $\beta = 1/(kT)$, where T is the temperature and k the Boltzmann constant, and $\hbar/k$ is still quite small, $\beta\hbar\omega \ll 1$ at finite temperature, so that this approximation makes sense. Then, go back to the exact expression for the entropy (7.57) and take into account that, in a neighborhood of zero, one has that $\coth x = 1/x + o(x)$ and $\sinh x = x + o(x^3)$. Taking $x = \beta\hbar\omega/2$ and first-order approximations for $\coth x$ and $\sinh x$, it results that (7.56) becomes (7.76). Therefore, the approximation to the value of the canonical entropy for the harmonic oscillator using path integration over coherent states makes sense.

7.4   A Discussion on the Definition of the Entropy for Quantum Non-relativistic Decaying Systems

A proper definition of the entropy for quantum non-relativistic decaying systems or resonances is still an open problem of which we want to introduce here some approximations. As for the case of the harmonic oscillator previously studied, one possibility would have been the introduction of a canonical entropy using formula (7.55). For this purpose, we first need either a basis in the space of decaying states in order to generalize the formula (7.56) or an approach including some sort of decaying coherent states. Another problem would be the choice of a proper model for resonances, although this can be easily solved with the use of the Friedrichs model that reproduces all the features of quantum resonances.

The Friedrichs model has been sketched in the present chapter. We have seen that $\{|\psi^D\rangle, |\omega_+\rangle\}$ is a sort of generalized basis including the Gamow state $|\psi^G\rangle$. Then, a first attempt to define the partition function Z following the philosophy of (7.56) could have been:

$$Z = \operatorname{Tr} e^{-\beta H} = \langle \psi^D | e^{-\beta H} | \psi^D \rangle + \int_0^\infty \langle \omega_+ | e^{-\beta H} | \omega_+ \rangle \, d\omega$$

$$= e^{-\beta z_R} \langle \psi^D | \psi^D \rangle + \int_0^\infty e^{-\beta \omega} \langle \omega_+ | \omega_+ \rangle \, d\omega \,. \qquad (7.77)$$

In order to obtain an explicit expression for $\langle \psi^D | \psi^D \rangle$, one should use (7.22) in (7.28) and (7.30). This would give

$$\langle \psi^D | \psi^D \rangle = 1 + \int_0^\infty \frac{\lambda^2 \, f^2(\omega)}{(z_R^* - \omega + i0)(z_R - \omega - i0)} \, d\omega \,. \qquad (7.78)$$

The integral (7.78) represents the action of a distribution on the function $f^2(\omega)$. This distribution is the product of $(z_R^* - \omega + i0)^{-1}$ times $(z_R - \omega - i0)^{-1}$, a product which is not defined on a distributional sense. Taking into account (7.37), a similar problem arises with $\langle \omega_+ | \omega_+ \rangle$. Therefore, this way seems to be non-practicable. The same problem appears with the use of the basis $\{|\psi^G\rangle, |\omega_-\rangle\}$. A different approach is in order here.

We may try to introduce at least an approximate notion of the canonical entropy for a system with a decaying state by using an analogy with the procedure used to approximate the canonical entropy for the harmonic oscillator. Thus, as in the case of the harmonic oscillator, we define the coherent state $|\alpha\rangle$ and its bra $\langle\alpha|$, as

$$|\alpha\rangle := \exp\{\alpha \, A_{IN}^\dagger\} |0\rangle \,, \qquad (7.79)$$

$$\langle\alpha| := \langle 0| \exp\{\alpha^* A_{OUT}\} \,, \qquad (7.80)$$

where $|0\rangle$ is a vacuum state, $A_{IN}^\dagger$ and A_{OUT} have been defined in (7.43) and (7.44), respectively. The vacuum state must describe a quantum system without decaying states. Making use of the commutation relations (7.50), we can prove that these coherent states $|\alpha\rangle$, $\alpha \in \mathbb{C}$, where $\mathbb{C}$ is the field of complex numbers, satisfy the expected properties for coherent states. In particular, we have that

$$A_{OUT} |\alpha\rangle = \alpha |\alpha\rangle, \quad \langle\alpha| A_{IN}^\dagger = \alpha \langle\alpha|, \quad \int_{\mathbb{C}} \frac{d^2\alpha}{\pi} |\alpha\rangle\langle\alpha| = 1, \quad d^2\alpha = d(\text{Real}\,\alpha)\, d(\text{Im}\,\alpha).$$
(7.81)

Then, we proceed to calculate the partition function as

$$Z := \text{Tr}\, e^{-\beta H} = \int_{\mathbb{C}} \frac{d^2\alpha}{\pi} \langle\alpha| e^{-\beta H} |\alpha\rangle = \int_{\mathbb{C}} \frac{d^2\alpha}{\pi} \rho(\alpha, \alpha).$$
(7.82)

For $\rho(\alpha, \alpha') = \langle\alpha| e^{-\beta H} |\alpha'\rangle$, we obtain the same expression as in (7.60). The explicit calculation is performed similarly as in the previous case. First, note that

$$H_+(\alpha, \alpha') = z_R\, \alpha^* \alpha'.$$
(7.83)

Then, we proceed to the integration as before, dropping the exponential terms causing divergencies. The final result is

$$\rho(\alpha, \alpha') = \exp\{-\beta z_R\, \alpha^* \alpha'\}.$$
(7.84)

Then, we use (7.83) in (7.82). This gives

$$Z = \frac{1}{\pi} \int_{-\infty}^{\infty} dx\, e^{-\beta z_R x^2} \int_{-\infty}^{\infty} dy\, e^{-\beta z_R y^2} = \frac{1}{\beta z_R}.$$
(7.85)

The final step uses (7.55) with (7.85) so as to give the desired result ($z_R = E_R - i\Gamma/2$):

$$S = k \left[1 - \ln\left(\beta \sqrt{E_R^2 + \frac{\Gamma^2}{4}} \right) - i \arctan\left(\frac{\Gamma}{2E_R} \right) \right].$$
(7.86)

In (7.86), we have taken the principal branch of $\log z$. The result given in (7.86) reminds the case of the harmonic oscillator, (7.76), except for the presence of an imaginary term. If $\Gamma \longmapsto 0$, both results do indeed coincide, after replacing $\hbar\omega$ by E_R. The presence of a complex entropy, for the case of Gamow vectors, requires some interpretation on the meaning of its imaginary part. The situation is quite similar to the existence of complex energy for decaying states, where the imaginary part is interpreted as the inverse of the half life.

Note that the resonance in the Friedrichs model is caused by the interaction of the system with the background, which plays the role of the thermodynamical bath. Then, we suggest that the real part of the entropy (7.86) is the entropy of the system and that the imaginary part of it is the entropy transferred from the system to the background. Should the thermodynamical entropy be identified with the modulus of (7.86), one concludes that the total entropy for a decaying state is bigger than the entropy of a stable system.

7.4.1 A Tentative for a More Accurate Approximation

Formula (7.86) is an approximation to the entropy of a quantum decaying state based on path integration over coherent states, a technique that has been previously shown how it works in the case of the one dimensional harmonic oscillator. The impression is that, as it happened for the harmonic oscillator, it may be an approximation of an exact formula. Here and using another approach, we attempt to provide a more accurate approximation for the entropy of a quantum decaying state.

We begin with an expression for the canonical density associated to a decaying system as

$$\rho = \frac{1}{Z} e^{-\beta H} = \frac{e^{-\beta H}}{\text{Tr}\,[e^{-\beta H}]}. \tag{7.87}$$

The Hamiltonian H should be written in a spectral decomposition including a dependence on operators representing the presence of decaying states. Here, we shall use the spectral decomposition for H in which the leading term contains the creation, $A_{IN}^{\dagger}$ and annihilation A_{OUT} operators for the Gamow vector $|\psi^D\rangle$. This Gamow vector is the representation of a decaying state.

In order to find the canonical entropy using (7.55), let us proceed as follows: Let us take the derivative with respect to β of $Z = \text{Tr}\,[e^{-\beta H}]$. This gives

$$-\frac{\partial Z}{\partial \beta} = \text{Tr}\,[e^{-\beta H}\, H]. \tag{7.88}$$

Then, use the expressions (7.53) and (7.54) for the Hamiltonian in (7.88). This gives

$$-\frac{\partial Z}{\partial \beta} = z_R \,\text{Tr}\,[e^{-\beta H}\, A_{IN}^{\dagger} A_{OUT}] + \int_0^{\infty} d\omega\, \omega\, \frac{\eta^+(\omega)}{\eta^-(\omega)}\, \text{Tr}\,[e^{-\beta H}\, B_{\omega\,IN}^{\dagger}\, B_{\omega\,OUT}]. \tag{7.89}$$

Let us consider the first trace in (7.89), where we have replaced $e^{-\beta H}$ by ρ as in (7.87). This just comes after multiplication by the constant Z^{-1}. Then, after the use of commutation relations (7.50), we obtain

$$\text{Tr}\,[\rho\, A_{IN}^{\dagger} A_{OUT}] = \text{Tr}\,[\rho(-1 + A_{OUT} A_{IN}^{\dagger})]. \tag{7.90}$$

Due to the definition (7.87) of ρ, its trace is equal to one, so that (7.90) must be equal to

$$-1 + \text{Tr}\,[\rho\, A_{OUT} A_{IN}^{\dagger}] = -1 + \text{Tr}\,[A_{IN}^{\dagger}\, \rho\, A_{OUT}], \tag{7.91}$$

where we have taken into account in (7.91) the cyclic property of the trace.

Next, let us introduce the following notation: Let τ be an arbitrary real number. For any operator O, we define its transformation $O(\tau)$ as

$$O(\tau) = e^{\tau H}\, O\, e^{-\tau H}. \tag{7.92}$$

In this presentation, O will be either $A_{IN}^\dagger$ or A_{OUT}. For $A_{IN}^\dagger$, we have that

$$[H, A_{IN}^\dagger(\tau)] = e^{\tau H}[H, A_{IN}^\dagger]e^{-\tau H} = \frac{\partial}{\partial \tau} A_{IN}^\dagger . \qquad (7.93)$$

Take the following commutator, in which we have used that all "A" operators commute with all "B" operators:

$$[H, A_{IN}^\dagger] = z_R(A_{IN}^\dagger A_{OUT} A_{IN}^\dagger - A_{IN}^\dagger A_{IN}^\dagger A_{OUT}) = z_R A_{IN}^\dagger [A_{OUT}, A_{IN}^\dagger] = z_R A_{IN}^\dagger , \qquad (7.94)$$

where we have used the first commutator in (7.50). Then, it is obvious that (7.93) and (7.94) yield to

$$\frac{\partial}{\partial \tau} A_{IN}^\dagger(\tau) = e^{\tau H}(z_R A_{IN}^\dagger) e^{-\beta H} = z_R A_{IN}^\dagger(\tau) . \qquad (7.95)$$

A similar result may be obtained for A_{OUT}, namely,

$$\frac{\partial}{\partial \tau} A_{OUT}(\tau) = -z_R A_{OUT}(\tau) . \qquad (7.96)$$

Integration of (7.95) and (7.96) is trivial and gives

$$A_{IN}^\dagger(\tau) = e^{\tau H} A_{IN}^\dagger ; \qquad A_{OUT}(\tau) = e^{-\tau H} A_{OUT} . \qquad (7.97)$$

The first equation in (7.97) can be written as

$$e^{\tau H} A_{IN}^\dagger e^{-\tau H} = e^{\tau z_R} A_{IN}^\dagger \implies A_{IN}^\dagger e^{-\tau H} = e^{\tau z_R} e^{-\tau H} A_{IN}^\dagger . \qquad (7.98)$$

Then, take the expression for the trace in the right-hand side of (7.91). This (7.98) and the choice $\tau = \beta$ give

$$\text{Tr}\,[A_{IN}^\dagger \rho A_{OUT}] = \frac{1}{Z} \text{Tr}\,(A_{IN}^\dagger e^{-\beta H} A_{OUT})$$

$$= \frac{e^{\beta z_R}}{Z}(e^{-\beta H} A_{IN}^\dagger A_{OUT}) = e^{\beta z_R} \text{Tr}\,[\rho A_{IN}^\dagger A_{OUT}] . \qquad (7.99)$$

From (7.90), (7.91) and (7.99), we conclude that

$$\text{Tr}\,[\rho A_{IN}^\dagger A_{OUT}] = -1 + e^{\beta z_R} \text{Tr}\,[\rho A_{IN}^\dagger A_{OUT}]$$

$$\implies \text{Tr}\,[\rho A_{IN}^\dagger A_{OUT}] = \frac{1}{e^{\beta z_R} - 1} . \qquad (7.100)$$

The use of the commutation relation in the right-hand side of (7.50) gives the following result:

$$\text{Tr}\,[e^{\beta H} B_{\omega\,IN}^\dagger B_{\omega'\,OUT}] = \frac{\eta^-(\omega)}{\eta^+(\omega)} \frac{1}{e^{\beta\omega}} \delta(\omega - \omega') . \qquad (7.101)$$

We observe that (7.101) depends on the continuous part (background) only and not on the value of z_R of the Gamow sector of the spectrum. This term is not defined (or it is infinite) for $\omega = \omega'$, but we can always drop it from the trace (7.89) since the background term is separated from the Gamow sector. Then, in the calculation of the partition function for the canonical Gamow state, we keep only the finite contributions to the trace (7.89). This is again some sort of regularization, where one gets rid of the contribution of the background and keeps that of the Gamow state only. With these ideas in mind, (7.89) yields

$$-\frac{1}{Z}\frac{\partial Z}{\partial \beta} = \frac{z_R}{e^{\beta z_R} - 1}. \tag{7.102}$$

Integration of (7.102) is trivial and gives

$$\log Z = -z_R \int \frac{d\beta}{e^{\beta z_R} - 1} = \beta z_R - \log(e^{\beta z_R} - 1). \tag{7.103}$$

Finally, we carry these results to (7.55) so as to obtain an approximation for the canonical entropy of a quantum decaying system, which is

$$S = k\left(\frac{\beta z_R e^{\beta z_R}}{e^{\beta z_R} - 1} - \log(e^{\beta z_R} - 1)\right). \tag{7.104}$$

We want to show that from (7.104) we may arrive to (7.86) after some approximations. First, note that by expanding $e^{\beta z_R} \approx 1 + \beta z_R + \dots$, the first term in parenthesis in (7.104) yields

$$\frac{\beta z_R e^{\beta z_R}}{e^{\beta z_R} - 1} \approx \frac{\beta z_R(1 + \dots)}{(1 + \beta z_R) - 1} \approx 1, \tag{7.105}$$

while the same for the second term reads

$$\log(e^{\beta z_R} - 1) \approx \log(\beta z_R). \tag{7.106}$$

Then, if $z = x + iy$ and use $\log z = \ln\sqrt{x^2 + y^2} + i\arctan(y/x)$, writing $z = \beta z_R = \beta(E_R - i\Gamma/2)$, we have that the right-hand side of (7.106) is equal to

$$\log(\beta z_R) = \ln\left(\beta\sqrt{E_R^2 + \frac{\Gamma^2}{2E_R}}\right) + i\arctan\left(\frac{\Gamma}{2E_R}\right). \tag{7.107}$$

The conclusion is that (7.86) may be viewed as an approximation of (7.107). Observe that we have arrived at the same type of approximation using two different methods, a fact which enforces the validity of the adopted procedure.

Note that, we have not made a specific definition of the trace in our discussion to obtain (7.104). In fact, for the derivation of (7.104), we have just assumed that the trace of a product of operators is invariant under cyclic permutation of the operators. The fact that our result is independent of the notion of trace allows us to circumvent the problems associated to the definition of norms in the presence of Gamow states.

7.5   Interaction Between a Gamow State and a Fermion

Next, we study an interesting case, that of the Gamow state interacting with a fermion, in order to get an expression of the entropy for the resulting unstable system. We assume that the total Hamiltonian is written as

$$H = H_F + \varepsilon \, |i\rangle\langle i| + H_f + \int_0^\infty V(\omega) \, [|i\rangle\langle i, \omega| + |i, \omega\rangle\langle i|] \, d\omega \,, \qquad (7.108)$$

where

- H_F is the Hamiltonian for the standard Friedrichs model written in the language of state vectors (see Sect. 8.2):

$$H_F := \omega_0 \, |\omega_0\rangle\langle\omega_0| + \lambda \int_0^\infty f(\omega) \, [|\omega\rangle\langle\omega_0| + |\omega_0\rangle\langle\omega|] \, d\omega + \int_0^\infty \omega \, |\omega\rangle\langle\omega| \, d\omega \,.$$
$$(7.109)$$

- $|i\rangle$ is the state of a free fermion.
- $H_f := h \, |i, \omega_0\rangle\langle i| + h^* \, |i\rangle\langle i, \omega_0|$ represents the interaction between the fermion $|i\rangle$ and the discrete boson in the Friedrichs model $|\omega_0\rangle$, h being a complex constant.
- $V(\omega)$ is the form factor for the interaction between the fermion and the external field.
- We use the notation $|i, \omega_0\rangle = |i\rangle \otimes |\omega_0\rangle$ and $|i, \omega\rangle = |i\rangle \otimes |\omega\rangle$, where $\otimes$ means tensor product. This means that we are assuming the existence of an interaction between the fermion and the boson field corresponding to the basic Friedrichs model.
- The functions $f(\omega)$ and $V(\omega)$ are admitted to have suitable analytic continuations on the complex plane.

This model exhibits at least one resonance. If again, we denote by $|\psi^G\rangle$ and $|\psi^D\rangle$ the eigenvectors of H with respective eigenvalues $z_R = E_R - i\Gamma/2$ and $z_R^* = E_R + i\Gamma/2$ (Gamow vectors), one may prove that H admits a spectral decomposition in terms of these Gamow vectors in the following form:

$$H = \varepsilon \, |i\rangle\langle i| + z_R \, |\psi^D\rangle\langle\psi^G| + \Lambda(z_R) \, [|i, \psi^D\rangle\langle i| + |i\rangle\langle i, \psi^G|] + B_G \,. \qquad (7.110)$$

Here, B_G denotes a background integral. We discard this term as we have done in previous sections, because we are mainly interested on the effect of the resonance. As before, $|i, \psi^D\rangle = |i\rangle \otimes |\psi^D\rangle$, etc. The value of the strength function also called the vertex function $\Lambda(z_R)$ can also be calculated and gives

$$\Lambda(z_R) = \lambda \, \mathrm{PV} \int_0^\infty \frac{f(\omega) \, V(\omega)}{z_R - \omega} \, d\omega - i\pi\lambda \, f(z_R) \, V(z_R) \,, \qquad (7.111)$$

where PV means the Cauchy principal value.

If we denote by $b^\dagger$ and b the creation and annihilation operators for the fermion, respectively, the relevant part of the Hamiltonian (without the background) can be written in the second quantization language as

$$H = \varepsilon\, b^\dagger b + z_R\, A_{IN}^\dagger\, A_{OUT} + \Lambda(z_R)\, b^\dagger b [A_{IN}^\dagger + A_{OUT}]. \qquad (7.112)$$

The properties of fermion operators and the independence of the fermion and boson sectors give the following relations:

$$\{b, b^\dagger\} = 1, \qquad [A_{OUT}, A_{IN}^\dagger] = 1, \qquad (7.113)$$

and all other commutators vanish.

Equation (7.112) comes up easily from the expansion of (7.110). One may show that the action of the Hamiltonian (7.110) (or (7.112)) produces no just one, but two distinct resonances. In order to obtain them, we have to solve the eigenvalue equation $H|\phi\rangle = z|\phi\rangle$ for z complex, where $|\phi\rangle$ is expressed in terms of the vectors $|i, 0\rangle$, $|i, \psi^D\rangle$, the so-called Berggren basis:

$$|\phi\rangle = \alpha\,|i, 0\rangle + \delta\,|i, \psi^D\rangle, \qquad (7.114)$$

where the zero in $|i, 0\rangle$ denotes the absence of the Gamow vector $|\psi^G\rangle$. In the two-dimensional space spanned by the vectors $|i, 0\rangle$ and $|i, \psi^D\rangle$, we may assume that these vectors are orthonormal (they are obviously linearly independent). The restriction of the Hamiltonian to the two-dimensional space spanned by this basis is given by

$$\begin{pmatrix} \varepsilon & \Lambda(z_R) \\ \Lambda(z_R) & \varepsilon + z_R \end{pmatrix}. \qquad (7.115)$$

The eigenvalues of this matrix are the complex energies

$$E_\pm = \varepsilon + \frac{z_R}{2} \pm \frac{1}{2}\sqrt{z_R^2 + 4\,\Lambda^2(z_R)}. \qquad (7.116)$$

In the sequel, we separate the unperturbed part, H_0, of the interacting Hamiltonian H from the interaction V in the following form:

$$H = H_0 + V, \qquad H_0 = \varepsilon\, b^\dagger b + z_R\, A_{IN}^\dagger\, A_{OUT}, \qquad (7.117)$$

$$V = \Lambda(z_R)\, b^\dagger b\, [A_{IN}^\dagger + A_{OUT}]. \qquad (7.118)$$

Note that, V is linear in the coupling constant of the original Friedrichs model, λ, due to the particular form of the vertex function $\Lambda(z_R)$ given in (7.111).

Then, the Gamow states for the interacting system can be easily calculated. For E_+, it has the following form:

$$|\phi_+\rangle = \alpha_+ |i, 0\rangle + \delta_+ |i, \psi^D\rangle$$

$$= \frac{1}{\Delta} \left\{ 2\Lambda(z_R) |i, 0\rangle + \left(z_R + \sqrt{z_R^2 + 4\Lambda^2(z_R)} \right) |i, \psi^D\rangle \right\} . \qquad (7.119)$$

For E_-:

$$|\phi_-\rangle = \alpha_- |i, 0\rangle + \delta_- |i, \psi^D\rangle$$

$$= \frac{1}{\Delta} \left\{ - \left(z_R + \sqrt{z_R^2 + 4\Lambda^2(z_R)} \right) |i, 0\rangle + 2\Lambda(z_R) |i, \psi^D\rangle \right\} . \qquad (7.120)$$

Here Δ is the trivial normalization constant that results assuming the orthonormality of $|i, 0\rangle$ and $|i, \psi^D\rangle$. Note that, under this point of view, Gamow vectors (7.119) and (7.120) are not orthogonal, since $\Lambda(z_R)$ is complex and therefore matrix (7.115) is not Hermitian.

We may evaluate the canonical entropy as we did in the previous case. Let us write the partition function as

$$Z = \text{Tr}\, e^{-\beta H(\lambda)} = e^{-\beta \Omega(\lambda)} . \qquad (7.121)$$

Equation (7.121) defines $\Omega(\lambda)$. We may write the explicit dependence of the interacting Hamiltonian H with λ just by writing $H = H_0 + \lambda W$. If we derive with respect to λ the first identity in (7.121), we have

$$\frac{\partial Z}{\partial \lambda} = -\beta \, \text{Tr}\, [e^{-\beta H(\lambda)} W] . \qquad (7.122)$$

If we define the average, $\langle W \rangle$ of W as

$$\langle W \rangle = \frac{\text{Tr}\, [e^{-\beta H(\lambda)}]}{Z} , \qquad (7.123)$$

then (7.122) is equal to

$$\frac{\partial Z}{\partial \lambda} = -\frac{\beta}{\lambda} e^{-\beta \Omega(\lambda)} \langle \lambda W \rangle . \qquad (7.124)$$

Then, if we derive with respect to λ the second identity in (7.121), we have

$$\frac{\partial Z}{\partial \lambda} = -e^{-\beta \Omega(\lambda)} \beta \frac{\partial \Omega(\lambda)}{\partial \lambda} . \qquad (7.125)$$

If we compare (7.124) to (7.125), we conclude that

$$\frac{\partial \Omega(\lambda)}{\partial \lambda} = \frac{1}{\lambda} \langle \lambda W \rangle. \tag{7.126}$$

Integration of (7.126) gives

$$\Omega(\lambda) = \Omega_0 + \int_0^1 \frac{1}{\lambda} \langle \lambda W \rangle \, d\lambda, \tag{7.127}$$

where Ω_0 is defined by the relation

$$e^{-\beta \Omega_0} := \text{Tr} \left[e^{-\beta H_0} \right]. \tag{7.128}$$

Equation (7.127) may be regarded as a perturbative expansion of the statistical potential Ω in powers of λ within the interval $0 < \lambda < 1$.

The next step is to calculate $\langle W \rangle$ using the thermal propagator method as described in a previous chapter. Let us write

$$\langle W \rangle = \frac{1}{2} \sum_k \lim_{k \to k'} \lim_{\tau \to \tau'} \left(\frac{\partial}{\partial \tau} \right) G(k, k', \tau - \tau'), \tag{7.129}$$

where

$$G(k, k', \tau - \tau') = -\langle T_\tau [a_k(\tau) a_{k'}^\dagger(\tau')] \rangle \tag{7.130}$$

is the thermal propagator, where T_τ is the ordering operator with respect to τ. The operators a_k appearing in (7.130) stand for the annihilation operators b, with $k = 1$, and A_{OUT}, with $k = 2$, while the creation operators $a_k^\dagger$ stand for $b^\dagger$, with $k = 1$, and $A_{IN}^\dagger$, with $k = 2$. The brackets $\langle - \rangle$ represent traces with respect to the Berggren basis $\{|i, 0\rangle, |i, \psi^D\rangle\}$ excluding the background, or equivalently with respect to the basis given by the Gamow vectors $|\phi_\pm\rangle$, which are eigenvectors of H. In general, (7.130) can be written as

$$G(k, k', \tau) = -\sum_n \langle n| e^{-\tau H} e^{\tau H} a_k e^{-\beta H} a_{k'}^\dagger |n\rangle$$

$$= -\sum_{nn'} \left(e^{-\beta E_n} e^{\tau(E_n - E_{n'})} \right) \langle n|a_k|n'\rangle \langle n'|a_{k'}^\dagger|n\rangle, \tag{7.131}$$

where $n = 1, 2$ and $|1\rangle := |\phi_+\rangle$ and $|2\rangle := |\phi_-\rangle$. Then, use the form for $|\phi_\pm\rangle$ given in (7.119) and (7.120), take the limits in (7.129) and taking into account the following relation, which the reader may easily check,

$$\langle \phi_\pm| b |\phi_\pm\rangle = \langle \phi_\pm| b^\dagger |\phi_\pm\rangle = 0, \tag{7.132}$$

we have that

$$G(2,2,0) = -e^{-\beta E_+} \langle \phi_+|A_{OUT}|\phi_+\rangle \langle \phi_+|A_{IN}^\dagger|\phi_+\rangle - e^{-\beta E_+} \langle \phi_+|A_{OUT}|\phi_-\rangle \langle \phi_-|A_{IN}^\dagger|\phi_+\rangle$$
$$= -e^{-\beta E_-} \langle \phi_-|A_{OUT}|\phi_+\rangle \langle \phi_+|A_{IN}^\dagger|\phi_-\rangle - e^{-\beta E_-} \langle \phi_-|A_{OUT}|\phi_-\rangle \langle \phi_-|A_{IN}^\dagger|\phi_-\rangle .$$
$$(7.133)$$

Let us calculate the brackets included in (7.133). First, taking into account that $A_{OUT}|i,0\rangle = 0$ and $A_{OUT}|i,\psi^D\rangle = |i,0\rangle$ and the orthogonality of vectors in the Berggren basis, we have that

$$\langle \phi_+|A_{OUT}|\phi_+\rangle = \langle (\alpha_+^*\langle i,0| + \delta_+^*\langle i,\psi^D|)|A_{OUT}|(\alpha_+|i,0\rangle + \delta_+|i,\psi^D\rangle))$$
$$= \langle (\alpha_+^*\langle i,0| + \delta_+^*\langle i,\psi^D|)|A_{OUT}|(\delta_+|i,\psi^D\rangle)) = \alpha_+^*\delta_+ . \qquad (7.134)$$

The second bracket is

$$\langle \phi_+|A_{IN}^\dagger|\phi_+\rangle = \langle (\alpha_+^*\langle i,0| + \delta_+^*\langle i,\psi^D|)|A_{IN}^\dagger|(\alpha_+|i,0\rangle + \delta_+|i,\psi^D\rangle))$$
$$= \langle (\alpha_+^*\langle i,0| + \delta_+^*\langle i,\psi^D|)|_{IN}^\dagger|(\alpha_+|i,0\rangle)) = \alpha_+\delta_+^* . (7.135)$$

As a matter of fact, $A_{IN}^\dagger|i,\psi^G\rangle$ should not be equal to zero, but instead equal to a vector with two Gamows. Under this point of view, we assume that the scalar product of this vector by another vector having zero or one Gamow should vanish. We may calculate the remainder brackets in (7.133) following this procedure. The final result is

$$G(2,2,0) = -e^{-\beta E_+} A(\lambda) - e^{-\beta E_-} B(\lambda) , \qquad (7.136)$$

with

$$A(\lambda) = |\alpha_+|^2|\delta_+|^2 + \alpha_+^*\delta_-\delta_+^*\alpha_- , \qquad (7.137)$$
$$B(\lambda) = \alpha_-^*\delta_+\delta_-^*\alpha_+ + |\alpha_-|^2|\delta_-|^2 . \qquad (7.138)$$

Note that the functions $A(\lambda)$ and $B(\lambda)$ are functions of λ due to the dependence of $\Lambda(z_R)$ on λ. Also, $E_\pm$ depends on λ through its dependence on $\Lambda(z_R)$ and z_R.

We may readily show that $G(2,2,0)$ is the unique relevant term in the sum (7.129) and that (7.129) along the above discussion gives

$$\langle \lambda W \rangle = E_+ e^{-\beta E_+} A(\lambda) + E_- e^{-\beta E_-} B(\lambda) , \qquad (7.139)$$

so that (7.127) takes the following form:

$$\Omega(\lambda) = \Omega_0 + \int_0^1 \frac{1}{\lambda} \{E_+ e^{-\beta E_+} A(\lambda) + E_- e^{-\beta E_-} B(\lambda)\} d\lambda = \Omega_0 + \Omega_I .$$
$$(7.140)$$

Since $\Omega(\lambda)$ is a sum of two terms, the canonical entropy is also a sum of two terms:

$$S = -\beta k \left(1 - \beta \frac{\partial}{\partial \beta}\right) \Omega_0 - \beta k \left(1 - \beta \frac{\partial}{\partial \beta}\right) \Omega_I = S_0 + S_I. \qquad (7.141)$$

To determine Ω_0 and hence S_0, we may use the method explained in the previous section. It is a simple exercise to show the validity of the following commutation relations:

$$[\varepsilon\, b^\dagger b, b^\dagger] = \varepsilon\, b^\dagger, \qquad [\varepsilon\, b^\dagger b, b] = -\varepsilon\, b, \qquad (7.142)$$

and, consequently,

$$[H_0, b^\dagger] = \varepsilon\, b^\dagger, \qquad [H_0, b] = -\varepsilon\, b. \qquad (7.143)$$

Then, let us take

$$\rho_0 = \frac{e^{-\beta H_0}}{Z_0}, \quad \text{with} \quad Z_0 := \text{Tr}\,[e^{-\beta H_0}], \qquad (7.144)$$

which gives

$$-\frac{1}{Z_0}\frac{\partial Z_0}{\partial \beta} = \text{Tr}\,[\rho_0 H_0] = z_R\,\text{Tr}\,[\rho_0 A_{IN}^\dagger A_{OUT}] + \varepsilon\,\text{Tr}\,[\rho_0 b^\dagger b]$$

$$= \frac{z_R}{e^{\beta z_R} - 1} + \frac{\varepsilon}{e^{\beta \varepsilon} + 1}. \qquad (7.145)$$

A simple integration of (7.145) gives

$$Z_0 = \beta\,\Omega_0 - \frac{(z_R - \varepsilon)\beta}{2} + \log\cosh\frac{\beta\varepsilon}{2} - \log\sinh\frac{\beta z_R}{2}. \qquad (7.146)$$

Then, it is a simple operation to obtain the value of S_0, which is the contribution to the canonical entropy due to the statistical potential Ω_0. This value of S_0 is

$$S_0 = k\left(\log\cosh\frac{\beta\varepsilon}{2} - \log\sinh\frac{\beta z_R}{2} - \frac{\beta\varepsilon}{2}\tanh\frac{\beta\varepsilon}{2} - \frac{\beta z_R}{2}\coth\frac{\beta z_R}{2}\right). \qquad (7.147)$$

Observe that S_0 is complex, since it contains the term z_R.

Next, we may find an approximation for the contribution S_I to the entropy. This contribution corresponds to the second term in the right hand side of (7.140) and express it into series of powers in λ. For $E_\pm$ the series comes from (7.111) and (7.116), see expansion of z_R in terms of λ in (7.149). The functions $A(\lambda)$ and $B(\lambda)$ are quadratic to the first order, i.e., $A(\lambda) = A_0\,\lambda^2 = o(\lambda^3)$ and the same for $B(\lambda)$. In order to get an explicit expression for $A(\lambda)$, take (7.137) and use the definition of the coefficients $\alpha_\pm$ and $\delta_\pm$ given in (7.119) and (7.120). This gives

$$A(\lambda) = \frac{4\,|\Lambda(z_R)|^2\,|z_R + \sqrt{z_R + 4\,\Lambda(z_R)}|^2}{[4\,|\Lambda(z_R)|^2 + |z_R + \sqrt{z_R + 4\,\Lambda(z_R)}|^2]^2} = \frac{N}{D}, \qquad (7.148)$$

where N and D represent the numerator and the denominator of (7.148), respectively. Let us write the numerator N up to the lowest order in λ. After the definition of $\Lambda(z_R)$ in (7.111) and the explicit of z_R for the basic Friedrichs model, which is

$$z_R = \omega_0 + \lambda^2 \int_0^\infty \frac{|f(\omega)|^2}{\omega_0 - \omega + i0} \, d\omega + o(\lambda^4) \,, \tag{7.149}$$

we have that in the lowest order in λ,

$$N \approx 4\lambda^2 \left| PV \int_0^\infty \frac{f(\omega)\, V(\omega)}{\omega_0 - \omega} \, d\omega - i\pi f(\omega_0)\, V(\omega_0) \right| 4\omega_0^2 = 16 c_0\, \omega_0^2 \, \lambda^2 \,, \tag{7.150}$$

where PV stands for principal value and c_0 is defined in (7.150). The form factor for the Friedrichs model, a role in our case played by the product $f(\omega)\, V(\omega)$, cannot vanish at the point ω_0, otherwise there is no resonance, as proven in the literature (Exner), so that $f(\omega_0)\, V(\omega_0) \neq 0$. The conclusion is that N is, at the lowest order, quadratic in λ. With respect to D, we can say that $D = 16\omega_0^4 + o(\lambda^4)$, so that

$$\frac{1}{D} = \frac{1}{16\omega_0^4} + o(\lambda^4), \tag{7.151}$$

hence

$$\frac{N}{D} = A_0\, \lambda^2 + o(\lambda^3)\,, \quad \text{with} \quad A_0 = \frac{c_0}{\omega_0^2} \,. \tag{7.152}$$

This shows that $A(\lambda)$ is quadratic at the lowest order in λ. Analogously, we show the same result for $B(\lambda)$. The conclusion is that the expression under the integral sign in (7.140),

$$\frac{1}{\lambda} \{ E_+ e^{-\beta E_+} A(\lambda) + E_- e^{-\beta E_-} B(\lambda) \} = a_1 \lambda + a_2 \lambda^2 + \dots \,, \tag{7.153}$$

admits a term by term integration. The determination of the coefficients $a_1, a_2, \dots$ comes after the explicit expressions for $E_\pm$, $A(\lambda)$ and $B(\lambda)$, which are known. This could be lengthy and tedious, but can be done at least for the lowest terms. This gives an approximation for S_I.

7.6   Summary

As a conclusion, we have pointed out to some of the difficulties concerning the application of concepts of statistical mechanics to complex-energy vectors. We have presented a suitable alternative to the probabilistic description, by implementing a representation of the decaying vectors, obtained in the framework of the Friedrichs model, written them in terms of coherent states and by performing a path integration over these states to get the density matrix operator. The results are quite encouraging because, at the level of approximation used to calculate the density operator, we do not have to introduce some ad hoc notions like complex temperatures or treat independently real and imaginary entropies. We think that this is a promising first step toward a novel formulation of the statistical mechanics for decaying systems.

References

1. Kubo, R.: Statistical Mechanics. North-Holland, Amsterdam (1965)
2. Huang, K.: Statistical Mechanics. Wiley, New York (1963)
3. Reichl, L.: A Modern Course in Statistical Physics. Edward Arnold, London (1980)
4. Pathria, R.K.: Statistical Mechanics. Pergamon, Oxford (1996)
5. Tolman, R.C.: The Principles of Statistical Mechanics. Clarendon Press, Oxford (1979)
6. Mandl, F.: Statistical Physics. Wiley, Chichester (1971)
7. Reif, F.: Fundamentals of Statistical and Thermal Physics. McGraw-Hill, New York (1965)
8. Landau, L.D., Lifshitz, E.M.: Physique Statistique. Editions MIR, Moscow (1967)
9. Hill, T.L.: An Introduction to Statistical Thermodynamics. Dover, New York (1987)
10. Bogolubov, N.N., Bogolubov Jr., N.N.: Introduction to Quantum Statistical Mechanics. Gordon and Breach, New York (1992)
11. Ma, S.-K.: Statistical Mechanics. World Scientific, Singapore (1985)
12. Lifshitz, E.M., Landau, L., Pitaevskii, L.P.: Statistical Physics. Elsevier, Amsterdam (1980)
13. Kubo, R., Ichimura, H., Usui, T., Hashitsume, N.: Statistical Mechanics. North-Holland, Amsterdam (1990)
14. Keldysh, L.V.: Sov. Phys. JETP **20**, 1018 (1965)
15. Umezawa, H., Matsumoto, H., Tachiki, M.: Thermo Field Dynamics and Condensed Matter. North-Holland, Amsterdam (1982)
16. ter Haar, D.: Lectures on Selected Topics in Statistical Mechanics. Pergamon Press, Oxford (1977)
17. Abrikosov, A.A., Gor'kov, L.P., Dzyaloshinskii, I.E.: Methods of Quantum Filed Theory in Statistical Physics. Prentice-Hall, Englewood Cliffs (1963)
18. Thouless, D.J.: The Quantum Mechanics of Many Body Systems. Academic, New York (1961)
19. Ring, P., Schuck, P.: The Nuclear Many Body Problem. Springer, New York (1980)
20. Feynman, R.P., Hibbs, A.R.: Quantum Mechanics and Path Integrals. MacGraw-Hill Book Company, New York (1965)
21. Popov, V.N.: Functional Integrals in Quantum Field Theory and Statistical Physics D. Reidel Publishing Company, Dordrecht (1983)
22. Glimm, J., Jaffe, A.: Quantum Physics: A Functional Point of View. Springer, New York (1987)
23. Bohm, A., Gadella, M.: Dirac Kets, Gamow Vectors and Gelfand Triplets. Lectures Notes in Physics, vol. 348, Springer, Berlin (1989)
24. Gel'fand, I.M., Shilov, G.E.: Generalized Functions, vol. VI-VIII. Academic, New York (1964)
25. Civitarese, O., Gadella, M.: Physics and mathematical aspects of Gamow states. Phys. Rep. **396**(2) (2004)

© The Editor(s) (if applicable) and The Author(s), under exclusive license
to Springer Nature Switzerland AG 2020
O. Civitarese and M. Gadella, *Methods in Statistical Mechanics*, Lecture Notes
in Physics 974, https://doi.org/10.1007/978-3-030-53658-9

Index

O. Civitarese and M. Gadella, *Methods in Statistical Mechanics*, Lecture Notes
in Physics 974, https://doi.org/10.1007/978-3-030-53658-9

Printed in the United States
By Bookmasters